TEAPACK Users' Manual
Mathematical Subroutines for

NUMERICAL METHODS

R. L. Johnston
University of Toronto

John Wiley & Sons
New York • Chichester • Brisbane • Toronto • Singapore

Copyright © 1982 by John Wiley & Sons, Inc.

All Rights Reserved.

Reproduction or translation of any part of
this work beyond that permitted by Section
107 or 108 of the 1976 United States Copyright
Act without the permission of the copyright
owner is unlawful. Requests for permission
or further information should be addressed to
the Permissions Department, John Wiley & Sons, Inc.

ISBN 0 471 86443 9
Printed in the United States of America

10 9 8 7 6 5 4 3 2 1

TABLE OF CONTENTS

I   INTRODUCTION   1

II   NUMERICAL LINEAR ALGEBRA

    1. Solution of an n x n system Ax = b

       (i) General real

✓ DCOMP	– LU decomposition of A	5
✓ SOLV	– forward and back substitution	7
✓ IMPRV	– iterative improvement	8

       (ii) Real, symmetric, positive definite

DSPD	– Cholesky decomposition of A	11
SSPD	– forward and back substitution	12

       (iii) Real, symmetric, positive definite, banded

DSPDB	– Cholesky decomposition of A	14
SSPDB	– forward and back substitution	16

       (iv) Tridiagonal

TRIDG	– solution of a tridiagonal system	18

    2. Eigenvalues and eigenvectors

✓ GCEV(V)	– eigenvalues (and eigenvectors) of a general n x n complex matrix	19
✓ GREV(V)	– eigenvalues (and eigenvectors) of a general n x n real matrix	21
✓ SREV(V)	– eigenvalues (and eigenvectors) of a real n x n symmetric matrix	23
SVD	– singular value decomposition of an m x n real matrix	25

    3. Overdetermined linear system Ax = b – least squares solution

✓ LLSNE	– normal equations method	27
LLSMG	– modified Gram-Schmidt method, rank(A) = n	29
MGS	– modified Gram-Schmidt method, rank(A) $\leq$ n	31

4. Utility routines

   (i) Matrices

AMAT	– matrix addition	34
MMAT	– matrix multiplication	35
TRANS	– transpose of an m x n matrix	36
ATA	– forms the product Transpose(A)*A	37

   (ii) Vectors

DPROD	– dot (or inner) product of two vectors	38
AMXNM	– maximum norm of a vector	39

   (iii) Matrix storage modes

PKS	– symmetric : full to packed	40
UPKS	– symmetric : packed to full	41
PKSB	– symmetric, banded : full to packed	42
UPKSB	– symmetric, banded : packed to full	43

III INTERPOLATION AND APPROXIMATION

1. Interpolation

PITPN	– polynomial, Newton form	44
PWLIN	– piecewise linear polynomial	45
PHERM	– piecewise cubic Hermite	46
SPLN	– cubic spline, free boundary	47
SPLNF	– cubic spline, clamped boundary	48

2. Approximation

LSQCS	– least squares, cubic spline	49

3. Evaluation routines

EVNWT	– polynomial, Newton form	51
EPLIN	– piecewise linear polynomial	53
EPCUB	– piecewise cubic polynomial	55

IV NONLINEAR EQUATIONS

1. Solution of a single equation

   (i) General equations

BSECT	– bisection method	57
SECNT	– secant method	59
NWTN	– Newton-Raphson method	61
MULR	– Muller's method	63
ZERO	– Brent's algorithm	65

(ii) Polynomial equations

    LAGR   – Laguerre's method   67

(iii) Utility routines for polynomial deflation and evaluation

    LDF   – deflation by $(x-r)$, r real   70
              – evaluation at $x = r$
    CLDF   – deflation by $(x-r)$, r complex   71
              – evalution at $x = r$
    CQDF   – deflation by $(x-r)(x-\bar{r})$ 72

2. Solution of a nonlinear system

    BRYDN   – Broyden's method   75
    SNWTN   – Newton's method   77

V QUADRATURE

1. Algorithms

    QUAD   – quadrature formula   80
    AQUAD   – adaptive quadrature   82
    ROM   – Romberg extrapolation   85
    CROM   – cautious Romberg extrapolation   87
    QUAD2   – quadrature formula – 2 dimensions   90

2. Formulas

    NCQ   – Newton-Cotes   92
    GLEGQ   – Gauss-Legendre   93
    GLAGQ   – Gauss-Laguerre   94

VI ORDINARY DIFFERENTIAL EQUATIONS

1. Algorithm

    IVODE   – solution using a one-step formula   95

2. Formulas

(i) Explicit formulas

    RKF4   – Runge-Kutta-Fehlberg 4th-order   100
    EULER   – Euler's formula   102

(ii) Implicit formulas

    TZN   – trapezoidal rule with Newton iteration   104

TZNF	– same as TZN but with Jacobian fixed	107
⁄TZMN	– trapezoidal rule with modified Newton iteration – finite difference Jacobian	110
TZMNF	– same as TZMN but with approximate Jacobian fixed	113

(iii) Utility routines (called by TZN and TZNF)

    FFU   116
    JFFU   117

INTRODUCTION

TEAPACK - a TEAching PACKage - is a package of mathematical subroutines, written in Fortran, that is intended primarily for use as an educational tool. It was developed at the University of Toronto where it has been used to support numerical methods courses for engineering and science students.

One of the design goals for TEAPACK was to incorporate some features that would facilitate experimentation with algorithms and investigation into their behavior. In this way, the package provides a laboratory for learning about the design and performance of computational algorithms. One such feature is an option for obtaining the output of intermediate results in order to examine the behavior of an algorithm under various circumstances. Many of the TEAPACK routines provide this option in the form of an extra parameter in the calling sequence. Another feature is a mechanism for choosing the underlying numerical method to be used by an algorithm. This idea has been incorporated into the quadrature and differential equations subroutines.

The overall organization of the package, including the format for the documentation in this manual, is more or less typical for mathematical software packages. Hence, TEAPACK provides a realistic environment within which a student can learn how to use software effectively, that is, how to select the most appropriate subroutine(s) for solving a particular problem at hand and how to interpret the information returned by a routine.

In developing this package, the educational aspects were given priority over criteria such as efficiency and robustness which are very important in the design of so-called production software. In this regard, it should be emphasized that TEAPACK is intended to be used as an educational tool only. It is not a sophisticated production software package such as IMSL or NAG and should not be used as a substitute for such packages.

Following are some general remarks about the package and this manual.

1. Contents. TEAPACK covers the topics normally discussed in a first course on scientific computing. These are:

> Numerical Linear Algebra
> Interpolation and Approximation
> Solution of Nonlinear Equations
> Quadrature

Ordinary Differential Equations - Initial-Value Problems.

2. _Portability._ The TEAPACK routines are written in ANSI Standard Fortran so that any Fortran compiler will accept them. In addition, whenever a machine-dependent constant is required by a subroutine, it is generated automatically by the routine itself. For example, many of the routines make use of machine epsilon EPS, or unit round-off level, as a basis for checking for accuracy or convergence. These routines contain the following program segment:

```
      EPS = 1.0
    1 EPS = 0.5*EPS
      EPSP1 = EPS + 1.0
      IF (EPSP1 .GT. 1.0) GO TO 1
```

This code will generate an approximation for EPS that differs from the exact value by at most a factor of 2, which is close enough. In these ways, then, the routines have been made independent of the particular computing system being used.

3. _Documentation._ The documentation for each routine consists of:

   (i) a brief description of the algorithm used;
   (ii) a definition of each parameter in the calling sequence; and
   (iii) a sample calling program.

The definitions of the parameters are classified according to their use as either input (ON ENTRY) or output (ON RETURN) parameters. In this way, it is easy to discern what information is required by a subroutine and what is returned. Sometimes a parameter is used for both functions in which case it will appear in both classifications. Although not stated explicitly in the documentation, such dual definitions imply that the input information is overwritten by the routine. Hence, if this information is required subsequently in the user's program, a copy of it should be made prior to calling the routine in question.

The sample calling programs are intended to illustrate how the routine can be used. These programs adhere to the rules of standard Fortran except for the input and output statements, where a format-free convention was adopted. This was done for clarity in reading the programs. Since many Fortran dialects support format-free I/O, the programs can very likely be run without modification. If, however, the local Fortran compiler does not support this feature, it will be understood that the I/O statements in the sample programs have to be modified appropriately before the programs can be run.

4. <u>Precision</u>. With one or two exceptions (noted below), each algorithm in TEAPACK has been implemented in both single and double precision. The documentation in this manual is for the single precision versions only. In order to access the corresponding double precision version of a routine, it is only necessary to:

(i) prefix the name of the routine with a D; and
(ii) change the declarations of all REAL variables to DOUBLE PRECISION.

Otherwise, the documentation is the same as for the single precision version. For example, in order to solve an n x n linear system of equations in double precision, the sample calling program for DCOMP-SOLV should be altered to:

```
      DOUBLE PRECISION A(10,10),B(10)
      DIMENSION NPIV(10)
      READ (5,*) N,((A(I,J),J=1,N),B(I),I=1,N)
      CALL DDCOMP(10,N,A,NPIV,IND)
      IF (IND .EQ. 0) GO TO 1
          WRITE (6,*) 'MATRIX IS NUMERICALLY SINGULAR'
          STOP
    1 CALL DSOLV(10,N,A,NPIV,B)
      WRITE (6,*) (B(I),I=1,N)
      STOP
      END
```

The only routine that does not have a double precision version is IMPRV - iterative improvement in the solution of a linear system of equations. This is because extended precision arithmetic must be used for computing the residual vector in order to minimize the effects of cancellation. Hence, a double precision version of IMPRV would require at least triple precision arithmetic, which is not available in standard Fortran. We remark that some compilers, such as H-level Fortran, support quadruple precision in which case a double precision version of IMPRV can easily be created by appropriate modifications to the declaration statements.

The function subprogram DPROD - dot product of two vectors - uses double precision arithmetic to accumulate the sum. The corresponding double precision version DDPROD should therefore use extended precision. However, since most Fortran compilers do not support this feature, it has not been used in DDPROD. In this sense, then, DDPROD is not an exact analogue of DPROD.

Some of the calculations in the eigenvalue routines GCEV, GCEVV, GREV, GREVV, the rootfinding routines MULR and LAGR, and the polynomial deflation routines CLDF and CQDF are done in complex arithmetic. Hence, double precision versions of these routines require double precision complex arithmetic (including all related intrinsic functions such as CDABS, CDSQRT, DCMPLX, DREAL and DIMAG), which is not included in standard Fortran. However, since many Fortran dialects support this feature, the routines DGCEV, DGCEVV, DGREV, DGREVV, DMULR, DLAGR, DCLDF and DCQDF have been included in the package. It should be understood, however, that they can only be used if the local Fortran compiler supports double precision complex arithmetic.

5. <u>Availability.</u> A complete copy of the TEAPACK package is available, upon request, at a nominal cost to cover distribution expenses. For details, write to

> TEAPACK Coordinator,
> Department of Computer Science,
> University of Toronto,
> Toronto, Ont. M5S 1A7
> Canada.

6. <u>Acknowledgments.</u> I am grateful for the assistance, encouragement and constructive criticisms from colleagues and students during the development of this package. In particular, I would like to acknowledge the contributions of Clifford Addison, Steve Ho-Tai and Richard Pancer who worked on the implementation, verification and testing of the subroutines.

```
      SUBROUTINE DCOMP(NDIM,N,A,NPIV,IND)
```

THIS SUBROUTINE, DECOMPOSITION WITH PARTIAL PIVOTING, DOES GAUSSIAN ELIMINATION OR, EQUIVALENTLY, A TRIANGULAR (LU) FACTORIZATION OF THE N X N MATRIX STORED IN THE ARRAY A. AT COMPLETION, THE ARRAY A WILL CONTAIN THE UNIT LOWER TRIANGULAR MATRIX L OF MULTIPLIERS USED IN THE ELIMINATION AS WELL AS THE UPPER TRIANGULAR MATRIX U, THE RESULT OF THE ELIMINATION.

THE ROUTINE INCLUDES A TEST FOR NUMERICAL SINGULARITY. THE CRITERION USED TO DEFINE NUMERICAL SINGULARITY IS:

   IF A PIVOT ELEMENT IS EQUIVALENT TO ZERO RELATIVE TO THE LARGEST PIVOT TO DATE, THAT IS, IF

           PIVMAX + PIV = PIVMAX

THIS ROUTINE CAN BE USED IN CONJUNCTION WITH THE ROUTINE SOLV TO FIND THE SOLUTION OF AN N X N SYSTEM OF LINEAR EQUATIONS A*X = B.

CALLING SEQUENCE: CALL DCOMP(NDIM,N,A,NPIV,IND)

PARAMETERS:

ON ENTRY:

    NDIM   - INTEGER
                THE DECLARED ROW DIMENSION OF THE ARRAY A IN
                THE CALLING PROGRAM. NDIM MUST BE >= N.

    N   - INTEGER
                THE ORDER OF THE MATRIX STORED IN A.

    A   - REAL(NDIM,N)
                THE MATRIX TO BE DECOMPOSED.

ON RETURN:

    A   - THE LU FACTORIZATION OF THE MATRIX:
                    L(I,J) = A(I,J),   1 <= I < J,
                    U(I,J) = A(I,J),   J <= I <= N.

    NPIV   - INTEGER(N)
                THIS IS A RECORD OF THE PIVOTING STRATEGY.
                THE VALUE K OF NPIV(I), 1<=I<=N-1, INDICATES
                THAT ROWS I AND K WERE INTERCHANGED AT THE
                I-TH STAGE OF THE ELIMINATION.
                THE VALUE OF NPIV(N) WILL BE +1 IF THERE WERE
                AN EVEN NUMBER OF ROW INTERCHANGES DURING THE
                ELIMINATION AND WILL BE -1 IF THERE WERE AN

ODD NUMBER.

IND   - INTEGER
          INDICATES IF A IS NUMERICALLY SINGULAR OR NOT.
     * { = -1   A IS NUMERICALLY SINGULAR
         = 0   A IS NUMERICALLY NONSINGULAR.

SAMPLE CALLING PROGRAM:

   SEE EITHER THE ROUTINE SOLV OR IMPRV

```
      SUBROUTINE SOLV(NDIM,N,LU,NPIV,B)
```

THIS SUBROUTINE PERFORMS THE FORWARD AND BACKWARD SUBSTITUTION STEPS IN THE SOLUTION OF A SYSTEM OF LINEAR EQUATIONS A*X = B. IT ASSUMES THAT THE TRIANGULAR (OR LU) FACTORIZATION OF A HAS ALREADY BEEN COMPUTED BY THE ROUTINE DCOMP. IF THIS ROUTINE HAS INDICATED THAT A IS NUMERICALLY SINGULAR, THEN THE USE OF SOLV MAY PRODUCE AN OVERFLOW INDICATION.

CALLING SEQUENCE: CALL SOLV(NDIM,N,LU,NPIV,B)

PARAMETERS:

ON ENTRY:

    NDIM   - INTEGER
                  THE DECLARED ROW DIMENSION OF THE ARRAY LU IN THE CALLING PROGRAM. NDIM MUST >= N.

    N   - INTEGER
                  THE ORDER OF THE SYSTEM TO BE SOLVED.

    LU   - REAL(NDIM,N)
                  THE LU DECOMPOSITION OF THE COEFFICIENT MATRIX A.

    NPIV   - INTEGER(N)
                  A RECORD OF THE ROW INTERCHANGES MADE DURING THE LU DECOMPOSITION OF A.

    B   - REAL(N)
                  THE RIGHT HAND SIDE OF THE SYSTEM.

ON RETURN:

    B   - THE COMPUTED SOLUTION X OF THE SYSTEM.

SAMPLE CALLING PROGRAM:

```
      DIMENSION A(10,10),B(10),NPIV(10)
      READ (5,*) N,((A(I,J),J=1,N),B(I),I=1,N)
      CALL DCOMP(10,N,A,NPIV,IND)
      IF (IND .EQ. 0) GO TO 1
          WRITE (6,*) 'MATRIX IS NUMERICALLY SINGULAR'
          STOP
    1 CALL  SOLV(10,N,A,NPIV,B)
      WRITE (6,*) (B(I),I=1,N)
      STOP
      END
```

SUBROUTINE IMPRV(NDIM,N,A,LU,NPIV,X,Z,R)

GIVEN AN APPROXIMATE SOLUTION FOR AN N X N SYSTEM OF LINEAR EQUATIONS A*X = B, THIS SUBROUTINE CARRIES OUT ONE ITERATION OF THE ITERATIVE IMPROVEMENT PROCESS FOR COMPUTING A BETTER APPROXIMATE SOLUTION.

IT IS ASSUMED THAT BOTH THE ORIGINAL MATRIX A AND ITS LU DECOMPOSITION AS WELL AS THE RIGHT HAND SIDE VECTOR B ARE AVAILABLE.

THIS SUBROUTINE CAN BE USED IN CONJUNCTION WITH THE SUBROUTINES DCOMP AND SOLV TO FIND THE SOLUTION OF A LINEAR SYSTEM OF EQUATIONS.

CALLING SEQUENCE: CALL IMPRV(NDIM,N,A,LU,NPIV,X,Z,R)

PARAMETERS:

ON ENTRY:

NDIM — INTEGER
THE DECLARED ROW DIMENSION OF THE ARRAYS A AND LU IN THE CALLING PROGRAM. NDIM MUST BE >= N.

N — INTEGER
THE ORDER OF THE SYSTEM TO BE SOLVED.

A — REAL(NDIM,N)
THE COEFFICIENT MATRIX A OF THE SYSTEM.

LU — REAL(NDIM,N)
THE LU DECOMPOSITION OF A.

NPIV — INTEGER(N)
A RECORD OF THE ROW INTERCHANGES MADE DURING THE DECOMPOSITION OF A.

X — REAL(N)
APPROXIMATE SOLUTION OF THE SYSTEM.

Z — REAL(N)
ON THE FIRST CALL, IT HOLDS THE INITIAL APPROXIMATE SOLUTION. ON SUBSEQUENT CALLS, IT HOLDS THE CORRECTION FOR THE PREVIOUS APPROXIMATE SOLUTION, I.E., THE CORRECTION Z PROVIDED ON RETURN FROM THE PREVIOUS CALL.

R — REAL(N)
ON THE FIRST CALL, IT HOLDS THE RIGHT HAND SIDE OF THE ORIGINAL SYSTEM. ON SUBSEQUENT

CALLS, IT HOLDS THE RESIDUAL FOR THE PREVIOUS APPROXIMATE SOLUTION, I.E., THE RESIDUAL R PROVIDED ON RETURN FROM THE PREVIOUS CALL.

ON RETURN:

    X   - THE IMPROVED APPROXIMATE SOLUTION.

    Z   - THE CORRECTION TO THE APPROXIMATE SOLUTION THAT WAS PROVIDED ON ENTRY.

    R   - THE RESIDUAL R = B - A*X, WHERE X IS THE APPROXIMATE SOLUTION ON ENTRY.

NOTE:

THE ROUTINE HAS BEEN DESIGNED TO FACILLIATE RE-CALLING TO DO ANOTHER ITERATION IN THAT THE VALUES OF X, Z AND R RETURNED BY THE PREVIOUS CALL ARE THOSE REQUIRED ON ENTRY BY THE RE-CALL. THEREFORE IT IS BEST NOT TO ALTER THE VALUES OF X, Z AND R BETWEEN CALLS TO IMPRV.

SAMPLE CALLING PROGRAM:

```
      DIMENSION A(10,10),B(10),R(10),Z(10),NPIV(10)
      REAL LU(10,10)
      READ (5,*) N,((A(I,J),J=1,N),B(I),I=1,N)
C
C     *************************
C     *
C     *  MAKE COPY OF A IN LU AND B IN R.
C     *
C     *************************
C
      DO 2 I=1,N
         R(I)=B(I)
         DO 1 J=1,N
            LU(I,J) = A(I,J)
1        CONTINUE
2     CONTINUE
      CALL DCOMP(10,N,LU,NPIV,IND)
      IF (IND .EQ. 0) GO TO 3
         WRITE (6,*) 'MATRIX IS NUMERICALLY SINGULAR'
         STOP
3     CALL SOLV(10,N,LU,NPIV,B)
      DO 4 I=1,N
         Z(I)=B(I)
4     CONTINUE
      CALL IMPRV(10,N,A,LU,NPIV,B,Z,R)
C
C     *************************
C     *
C     *  ESTIMATE THE CONDITION NUMBER OF THE MATRIX.
C     *  THE CONSTANT ALPH IS THE SMALLEST MACHINE
C     *  NUMBER SUCH THAT:
```

```
C     *
C     *                ALPH + 1.0 = ALPH
C     *
C     *   FOR EXAMPLE, IN THE FLOATING POINT NUMBER
C     *   SYSTEM F(16,6,L,U), ALPH = 16**6.
C     *
C     **************************
C
      RELCOR = SQRT(DPROD(N,Z,Z)/DPROD(N,B,B))
      CONDA = RELCOR * ALPH
      WRITE (6,*) 'ESTIMATE FOR COND(A) IS', CONDA
      ITER = 1
      IF (RELCOR .LE. 1.0E-5) GOTO 6
C
C     **************************
C     *
C     *   PERFORM AT MOST 4 MORE ITERATIONS UNTIL THE
C     *   APPROXIMATE SOLUTION IS CORRECT TO 5 SIGNIFICANT
C     *   FIGURES.
C     *
C     **************************
C
      DO 5 ITER = 2, 5
         CALL IMPRV(10,N,A,LU,NPIV,B,Z,R)
         RELCOR = SQRT(DPROD(N,Z,Z)/DPROD(N,B,B))
         IF (RELCOR .LE. 1.0E-5) GOTO 6
    5 CONTINUE
      WRITE (6,*) 'ITERATIVE IMPROVEMENT FAILED TO CONVERGE
     +            IN FIVE ITERATIONS'
      GOTO 7
    6 WRITE (6,*) 'ITERATIVE IMPROVEMENT CONVERGED AFTER',
     +            ITER, 'ITERATIONS'
    7 WRITE (6,*) 'FINAL APPROXIMATE SOLUTION AND CORRECTIONS ARE:'
      DO 8 I = 1, N
         WRITE (6,*) I, B(I), Z(I)
    8 CONTINUE
      STOP
      END
```

```
      SUBROUTINE DSPD(N,AP,IND)
```

THIS SUBROUTINE FACTORS A SYMMETRIC POSITIVE DEFINITE MATRIX A = TRANS(R)*R WHERE R IS AN UPPER TRIANGULAR MATRIX. THE CHOLESKY METHOD IS USED. IT IS ASSUMED THAT A IS STORED IN PACKED FORM (SEE DEFINITION OF THE ARRAY AP). THE SUBROUTINE PKS MAY BE USED TO PUT A IN PACKED FORM.

THIS ROUTINE CAN BE USED IN CONJUNCTION WITH THE ROUTINE SSPD TO FIND THE SOLUTION OF A LINEAR SYSTEM OF EQUATIONS A*X = B, WHERE A IS SYMMETRIC AND POSITIVE DEFINITE.

CALLING SEQUENCE: CALL DSPD(N,AP,IND)

PARAMETERS:

ON ENTRY:

- AP   - REAL(N*(N+1)/2)
            THE UPPER TRIANGLE (COLUMN BY COLUMN) OF THE SYMMETRIC MATRIX A. SPECIFICALLY,
                AP(IJ) = A(I,J), 1<=I<=J, 1<=J<=N,
            WHERE IJ = J*(J-1)/2 + I.

- N   - INTEGER
            THE ORDER OF THE MATRIX A.

ON RETURN:

- AP   - THE UPPER TRIANGLAR MATRIX R, STORED IN PACKED FORM, SUCH THAT A = TRANS(R)*R.
            IF IND .NE. 0, THE FACTORIZATION HAS NOT BEEN COMPLETED.

- IND   - INTEGER
            = 0 THE FACTORIZATION WAS SUCCESSFULLY COMPLETED,
            = K ERROR CONDITION. THE LEADING MINOR OF ORDER K IS NOT POSITIVE DEFINITE.

SAMPLE CALLING PROGRAM:

    SEE THE ROUTINE SSPD.

```
      SUBROUTINE SSPD(N,RP,B)
```

THIS SUBROUTINE FINDS THE SOLUTION OF THE LINEAR SYSTEM OF EQUATIONS A*X = B, WHERE A IS AN N*N SYMMETRIC, POSITIVE DEFINITE MATRIX. IT IS ASSUMED THAT A IS IN THE FACTORED FORM A = TRANS(R)*R AS MAY BE COMPUTED USING THE SUBROUTINE DSPD.

CALLING SEQUENCE: CALL SSPD(N,RP,B)

PARAMETERS:

ON ENTRY:

    N   - INTEGER
             THE ORDER OF THE MATRIX A.

    RP   - REAL(N*(N+1)/2)
             THE UPPER TRIANGULAR MATRIX, STORED IN PACKED FORM, SUCH THAT A = TRANS(R)*R.

    B   - REAL(N)
             THE RIGHT HAND SIDE OF THE SYSTEM.

ON RETURN:

    B   - REAL(N)
             THE SOLUTION VECTOR.

SAMPLE CALLING PROGRAM:

```
      DIMENSION AP(55),B(10)
C
C     **************************
C     *
C     *  INPUT  THE MATRIX A AND STORE IN PACKED FORM.  THE DATA
C     *  SHOULD BE ORDERED BY COLUMNS AS FOLLOWS:
C     *
C     *     N
C     *     A(1,1)  A(1,2)  A(2,2) . . .  A(1,N) ... A(N,N)
C     *     B(1)  B(2)  ...  B(N)
C     *
C     *  NOTE: IF THE FULL MATRIX A HAS BEEN INPUT, THE ROUTINE
C     *  PKS CAN BE USED TO CONVERT TO PACKED FORM.
C     *
C     **************************
C
      READ (5,*) N
      NS = N*(N+1)/2
      READ (5,*) (AP(I),I = 1,NS)
      READ (5,*) (B(I),I=1,N)
```

```
      CALL   DSPD(N,AP,IND)
      IF (IND .EQ. 0) GO TO 1
          WRITE (6,*) 'LEADING MINOR OF ORDER',IND,
     +                ' IS NOT POSITIVE DEFINITE'
          STOP
C
C     **************************
C     *
C     *   COMPUTE THE SOLUTION VECTOR.
C     *
C     **************************
C
    1 CALL   SSPD(N,AP,B)
      WRITE (6,*) (B(I),I=1,N)
      STOP
      END
```

           SUBROUTINE DSPDB(MDIM,M,N,AP,IND)

THIS SUBROUTINE DECOMPOSES A SYMMETRIC POSITIVE DEFINITE
BAND MATRIX A=TRANS(R)*R WHERE R IS AN UPPER TRIANGULAR BAND
MATRIX. THE CHOLESKY METHOD IS USED. IT IS ASSUMED THAT A
IS STORED IN PACKED FORM (SEE DEFINITION OF THE ARRAY AP).
THE SUBROUTINE PKBS MAY BE USED TO PUT A IN PACKED FORM.

THIS ROUTINE CAN BE USED IN CONJUNCTION WITH THE ROUTINE
SSPDB TO FIND THE SOLUTION OF A LINEAR SYSTEM OF EQUATIONS
A*X = B, WHERE A IS BAND, SYMMETRIC, AND POSITIVE DEFINITE.

CALLING SEQUENCE: CALL DSPDB(MDIM,M,N,AP,IND)

PARAMETERS:

ON ENTRY:

        MDIM   - INTEGER
                  THE DECLARED ROW DIMENSION OF THE ARRAY AP IN
                  THE CALLING PROGRAM. MDIM MUST BE >= M+1.

        M   - INTEGER
                  THE NUMBER OF DIAGONALS ABOVE THE MAIN
                  DIAGONAL OF THE MATRIX A. (THE BAND WIDTH OF
                  A IS 2*M+1).

        N   - INTEGER
                  THE ORDER OF THE MATRIX A.

        AP   - REAL(MDIM,N)
                  THE SYMMETRIC BAND MATRIX, STORED IN PACKED
                  FORM. THE DIAGONALS OF A ARE THE ROWS OF AP.
                  SPECIFICALLY, THE NONZERO ENTRIES OF AP ARE
                  GIVEN BY

                      AP(IJ,J) = A(I,J),   1<=I<=J<=I+M<=N,

                  WHERE IJ = M+1+I-J.

ON RETURN:

        AP   - THE (BAND) UPPER TRIANGULAR MATRIX R, STORED
                  IN PACKED FORM, IN THE DECOMPOSITION
                  A = TRANS(R)*R.

        IND   - INTEGER
                  = 0   FOR NORMAL RETURN,
                  = K   IF THE LEADING MINOR OF ORDER K IS NOT
                        POSITIVE DEFINITE.

SAMPLE CALLING PROGRAM:

SEE THE ROUTINE SSPDB.

```
      SUBROUTINE SSPDB(MDIM,M,N,RP,B)
```

THIS SUBROUTINE FINDS THE SOLUTION OF THE LINEAR SYSTEM OF EQUATIONS A*X = B, WHERE A IS AN N X N SYMMETRIC, POSITIVE DEFINITE, BAND MATRIX. IT IS ASSUMED THAT A IS IN THE FACTORED FORM A = TRANS(R)*R AS MAY BE COMPUTED USING THE SUBROUTINE DSPDB.

CALLING SEQUENCE: CALL SSPDB(MDIM,M,N,RP,B)

PARAMETERS:

ON ENTRY:

    MDIM   - INTEGER
                THE DECLARED ROW DIMENSION OF THE ARRAY RP IN THE CALLING PROGRAM. MDIM MUST BE >= M+1.

    M   - INTEGER
                THE NUMBER OF DIAGONALS ABOVE THE MAIN DIAGONAL OF THE MATRIX A. (THE BAND WIDTH OF A IS 2*M+1).

    N   - INTEGER
                THE ORDER OF THE MATRIX A.

    RP   - REAL(MDIM,N)
                THE (BAND) UPPER TRIANGULAR MATRIX, STORED IN PACKED FORM, SUCH THAT A = TRANS(R)*R.

    B   - REAL(N)
                THE RIGHT HAND SIDE OF THE SYSTEM.

ON RETURN:

    B   - THE SOLUTION VECTOR.

SAMPLE CALLING PROGRAM:

```
      DIMENSION AP(5,10),B(10)
C
C     *************************
C     *
C     *  WITH THIS EXAMPLE, ONLY THE ELEMENTS IN THE NONZERO
C     *  BAND OF A ARE INPUT.  THE DATA SHOULD BE ORDERED BY
C     *  DIAGONALS, STARTING WITH THE MAIN DIAGONAL, AS
C     *  FOLLOWS:
C     *
C     *      M  N
C     *      A(1,1)   A(2,2)   . . .   A(N,N)
C     *      A(1,2)   A(2,3)   ...   A(N-1,N)
```

```fortran
C     *           .
C     *            .
C     *             A(1,M+1)  ...  A(N-M,N)
C     *             B(1)  B(2) . . .  B(N)
C     *
C     *  NOTE:  IF THE FULL MATRIX A IS INPUT, THE  ROUTINE
C     *  PKSB CAN BE USED TO CONVERT TO PACKED FORM.
C     *
C     ************************
C
      READ (5,*) M,N
      MP1 = M + 1
      DO 10 KD = 1,MP1
         KDIAG = M + 2 - KD
         READ (5,*) (AP(KDIAG,IJ),IJ=KD,N)
   10 CONTINUE
      READ (5,*) (B(I),I=1,N)
C
      CALL DSPDB(5,M,N,AP,IND)
      IF (IND .EQ. 0) GO TO 30
          WRITE (6,*) 'LEADING MINOR OF ORDER',IND,
     +                ' IS NOT POSITIVE DEFINITE'
          STOP
   30 CALL SSPDB(5,M,N,AP,B)
      WRITE (6,*) (B(I),I=1,N)
      STOP
      END
```

```
      SUBROUTINE TRIDG(N,AD,B)
```

THIS SUBROUTINE FINDS THE SOLUTION OF THE LINEAR SYSTEM OF EQUATIONS A*X = B, WHERE A IS AN N X N TRIDIAGONAL MATRIX.

CALLING SEQUENCE: CALL TRIDG(N,AD,B)

PARAMETERS:

ON ENTRY:

    N   - INTEGER
                THE ORDER OF THE MATRIX A.

    AD   - REAL(3,N)
                THE TRIDIAGONAL BAND OF A:
                   AD(1,J) = A(J,J-1),   2<=J<=N;
                   AD(2,J) = A(J,J),   1<=J<=N;
                   AD(3,J) = A(J,J+1),   1<=J<=N-1.

    B   - REAL(N)
                THE RIGHT HAND SIDE OF THE SYSTEM.

ON RETURN

    B   - THE SOLUTION VECTOR.

SAMPLE CALLING PROGRAM:

```
      DIMENSION AD(3,10),B(10)
C
C     ***************
C     *
C     * INPUT THE TRIDIAGONAL BAND OF A AND THE VECTOR B.
C     *
C     ***************
C
      READ (5,*) N
      READ (5,*) (AD(1,J),J=2,N)
      READ (5,*) (AD(2,J),J=1,N)
      NM1 = N - 1
      READ (5,*) (AD(3,J),J=1,NM1)
      READ (5,*) (B(I),I=1,N)
      CALL TRIDG(N,AD,B)
      WRITE(6,*) (B(I),I=1,N)
      STOP
      END
```

```
      SUBROUTINE GCEV (NDIM,N,AR,AI,WR,WI,IERR)
              GCEVV(NDIM,N,AR,AI,WR,WI,ZR,ZI,IERR)
```

THE SUBROUTINE GCEV WILL FIND THE EIGENVALUES OF AN N X N GENERAL COMPLEX MATRIX.

THE SUBROUTINE GCEVV WILL FIND THE EIGENVALUES AND EIGENVECTORS OF AN N*N GENERAL COMPLEX MATRIX.

THESE ROUTINES ARE SLIGHTLY MODIFIED VERSIONS OF THE ROUTINES GENCV AND GENCVV FROM THE EISPACK PACKAGE. THE MODIFICATION IS THAT MACHINE-DEPENDENT CONSTANTS ARE GENERATED RATHER THAN ASSIGNED.
(SEE: MATRIX EIGENSYSTEM ROUTINES - EISPACK GUIDE
        BY B. SMITH, ET AL.)

CALLING SEQUENCE: CALL GCEV(NDIM,N,AR,AI,WR,WI,IERR)
                     CALL GCEVV(NDIM,N,AR,AI,WR,WI,ZR,ZI,IERR)

PARAMETERS:

ON ENTRY:

    NDIM   - INTEGER
              THE DECLARED ROW DIMENSION OF THE ARRAYS AR
              AND AI IN THE CALLING PROGRAM. NDIM MUST
              BE >= N.

    N   - INTEGER
              THE ORDER OF THE MATRIX FOR WHICH THE
              EIGENVALUES ARE SOUGHT.

    AR,AI   - REAL(NDIM,N)
              AR AND AI CONTAIN THE REAL AND IMAGINARY
              PARTS, RESPECTIVELY, OF THE GENERAL COMPLEX
              MATRIX OF ORDER N FOR WHICH THE EIGENVALUES
              ARE SOUGHT.

ON RETURN:

    AR,AI   - THE ORIGINAL MATRICES STORED IN AR AND AI ARE
              DESTROYED.

    WR,WI   - REAL(N)
              WR AND WI CONTAIN THE REAL AND IMAGINARY
              PARTS, RESPECTIVELY, OF THE EIGENVALUES OF A.

    ZR,ZI   - REAL(NDIM,N)
              THE COLUMNS OF ZR AND ZI CONTAIN THE REAL
              AND IMAGINARY PARTS, RESPECTIVELY, OF THE
              EIGENVECTORS OF A. THE EIGENVECTORS ARE NOT
              NORMALIZED.

IERR   - INTEGER
          AN ERROR COMPLETION CODE. THE NORMAL
          COMPLETION CODE IS ZERO. IF MORE THAN 30
          ITERATIONS ARE REQUIRED TO DETERMINE AN
          EIGENVALUE, GCEV(V) TERMINATES WITH IERR
          SET TO THE INDEX OF THE EIGENVALUE FOR
          WHICH THE FAILURE OCCURRED. THE EIGENVALUES
          IN THE WR AND WI ARRAYS SHOULD BE CORRECT FOR
          THE INDICES IERR+1, IERR+2, ..., N, BUT NO
          EIGENVECTORS ARE COMPUTED.

```
SUBROUTINE GREV (NDIM,N,WR,WI,IERR)
               GREVV(NDIM,N,WR,WI,Z,IERR)
```

THE SUBROUTINE GREV WILL FIND THE EIGENVALUES OF AN N X N GENERAL REAL MATRIX.

THE SUBROUTINE GREVV WILL FIND THE EIGENVALUES AND EIGENVECTORS OF AN N*N GENERAL REAL MATRIX.

THESE ROUTINES ARE SLIGHTLY MODIFIED VERSIONS OF THE ROUTINES GENRV AND GENRVV FROM THE EISPACK PACKAGE. THE MODIFICATION IS THAT MACHINE-DEPENDENT CONSTANTS ARE GENERATED RATHER THAN ASSIGNED.
(SEE: MATRIX EIGENSYSTEM ROUTINES - EISPACK GUIDE
        BY B. SMITH, ET AL.)

CALLING SEQUENCE: CALL GREV(NDIM,N,A,WR,WI,IERR)
                    CALL GREVV(NDIM,N,A,WR,WI,Z,IERR)

PARAMETERS:

ON ENTRY:

    NDIM   - INTEGER
              THE DECLARED ROW DIMENSION OF THE ARRAY A IN THE CALLING PROGRAM. NDIM MUST BE >= N.

    N   - INTEGER
              THE ORDER OF THE MATRIX FOR WHICH THE EIGENVALUES ARE SOUGHT.

    A   - REAL(NDIM,N)
              THE GENERAL REAL MATRIX OF ORDER N FOR WHICH THE EIGENVALUES ARE SOUGHT.

ON RETURN:

    A   - THE ORIGINAL MATRIX STORED IN A IS DESTROYED.

    WR,WI   - REAL(N)
              WR AND WI CONTAIN THE REAL AND IMAGINARY PARTS, RESPECTIVELY, OF THE EIGENVALUES OF A. THE EIGENVALUES ARE UNORDERED EXCEPT THAT COMPLEX CONJUGATE PAIRS OF EIGENVALUES APPEAR CONSECUTIVELY WITH THE ONE HAVING THE POSITIVE IMAGINARY PART APPEARING FIRST.

    Z   - REAL(NDIM,N)
              THE COLUMNS OF Z CONTAIN THE REAL AND IMAGINARY PARTS OF THE EIGENVECTORS OF A. IF THE J-TH EIGENVALUE IS REAL, THE J-TH COLUMN OF Z WILL CONTAIN ITS EIGENVECTOR. IF THE J-TH EIGENVALUE IS COMPLEX WITH POSITIVE

IMAGINARY PART, THE J-TH AND (J+1)-TH COLUMNS OF Z WILL CONTAIN THE REAL AND IMAGINARY PARTS, RESPECTIVELY, OF ITS EIGENVECTOR. THE CONJUGATE OF THIS VECTOR IS THE EIGENVECTOR FOR THE CONJUGATE EIGENVALUE. THE EIGENVECTORS ARE NOT NORMALIZED.

IERR   - INTEGER
AN ERROR COMPLETION CODE. THE NORMAL COMPLETION CODE IS ZERO. IF MORE THAN 30 ITERATIONS ARE REQUIRED TO DETERMINE AN EIGENVALUE, GREV(V) TERMINATES WITH IERR SET TO THE INDEX OF THE EIGENVALUE FOR WHICH THE FAILURE OCCURRED. THE EIGENVALUES IN THE WR AND WI ARRAYS SHOULD BE CORRECT FOR THE INDICES IERR+1, IERR+2, ..., N, BUT NO EIGENVECTORS ARE COMPUTED.

```
          SUBROUTINE SREV (NDIM,N,A,W,IERR)
                  SREVV(NDIM,N,A,W,Z,IERR)
```

THE SUBROUTINE SREV WILL FIND THE EIGENVALUES OF AN N X N REAL SYMMETRIC MATRIX.

THE SUBROUTINE SREVV WILL FIND THE EIGENVALUES AND EIGENVECTORS OF AN N*N REAL SYMMETRIC MATRIX.

THESE ROUTINES ARE SLIGHTLY MODIFIED VERSIONS OF THE ROUTINES SYMRV AND SYMRVV FROM THE EISPACK PACKAGE. THE MODIFICATION IS THAT MACHINE-DEPENDENT CONSTANTS ARE GENERATED RATHER THAN ASSIGNED.
(SEE: MATRIX EIGENSYSTEM ROUTINES - EISPACK GUIDE
      BY B. SMITH, ET AL.)

CALLING SEQUENCE: CALL SREV(NDIM,N,A,W,IERR)
                     CALL SREVV(NDIM,N,A,W,Z,IERR)

PARAMETERS:

ON ENTRY:

   NDIM   - INTEGER
               THE DECLARED ROW DIMENSION OF THE ARRAY A IN
               THE CALLING PROGRAM. NDIM MUST BE >= N.

   N   - INTEGER
               THE ORDER OF THE MATRIX FOR WHICH THE
               EIGENVALUES ARE SOUGHT.

   A   - REAL(NDIM,N)
               THE SYMMETRIC MATRIX OF ORDER N FOR WHICH THE
               EIGENVALUES ARE SOUGHT. ONLY THE FULL
               LOWER TRIANGLE OF THE MATRIX NEED BE
               SUPPLIED.

ON RETURN:

   A   - THE ORIGINAL MATRIX IN THE LOWER TRIANGLE IS
               OVER-WRITTEN BUT THE UPPER TRIANGLE IS UN-
               ALTERED.

   W   - REAL(N)
               THE EIGENVALUES OF A IN ASCENDING ORDER.

   Z   - REAL(NDIM,N)
               THE COLUMNS OF Z WILL CONTAIN THE EIGEN-
               VECTORS OF A.

   IERR   - INTEGER
               AN ERROR COMPLETION CODE. IF MORE THAN 30
               ITERATIONS ARE REQUIRED TO DETERMINE AN

EIGENVALUE, SREV(V) TERMINATES WITH IERR SET TO THE INDEX OF THE EIGENVALUE WHERE THE FAILURE OCCURRED. THE EIGENVALUES (AND EIGENVECTORS) IN THE W (AND Z) ARRAYS SHOULD BE CORRECT FOR INDICES 1,2,...,IERR-1, BUT THE EIGENVALUES ARE UNORDERED. IF ALL THE EIGENVALUES ARE DETERMINED WITHIN 30 ITERATIONS, IERR IS SET TO ZERO.

        SUBROUTINE SVD(MDIM,M,N,A,W,MATU,U,MATV,V,IERR,RV1)

THIS ROUTINE WILL COMPUTE THE SINGULAR VALUES AND THE
COMPLETE ORTHOGONAL DECOMPOSITION OF A REAL M X N MATRIX A.
A IS DECOMPOSED INTO

            U*DIAG(S)*TRANS(V),

WHERE U AND V, BOTH ORTHOGONAL, ARE OF ORDER M AND N
RESPECTIVELY; AND S IS AN M*N 'DIAGONAL' MATRIX WHOSE
DIAGONAL ELEMENTS ARE THE SINGULAR VALUES OF A, I.E., THE
NON-NEGATIVE SQUARE ROOTS OF THE EIGENVALUES OF TRANS(A)*A.

THIS ROUTINE IS TAKEN FROM THE EISPACK PACKAGE. IT HAS BEEN
SLIGHTLY MODIFIED IN THAT MACHINE-DEPENDENT CONSTANTS ARE
GENERATED RATHER THAN ASSIGNED.
(SEE: MATRIX EIGENSYSTEM ROUTINES - EISPACK GUIDE
       BY B. SMITH, ET AL.)

CALLING SEQUENCE: CALL SVD(MDIM,M,N,A,W,MATU,U,MATV,V,
                              IERR,RV1)

PARAMETERS:

ON ENTRY:

     MDIM   - INTEGER
                 THE DECLARED ROW DIMENSION OF THE ARRAYS A, U
                 AND V IN THE CALLING PROGRAM. MDIM MUST
                 BE >= M, N.

     M   - INTEGER
                 THE NUMBER OF ROWS IN THE MATRICES A   AND   U.

     N   - INTEGER
                 THE NUMBER OF COLUMNS OF THE MATRICES A AND U
                 AND   THE   ORDER OF   THE   SQUARE   MATRIX   V.

     A   - REAL(MDIM,N)
                 THE MATRIX TO BE DECOMPOSED.

     MATU   - LOGICAL
                 SET TRUE IF THE MATRIX U IS DESIRED AND SET
                 FALSE OTHERWISE.

     MATV   - LOGICAL
                 SET TRUE IF THE MATRIX V IS DESIRED AND SET
                 FALSE OTHERWISE.

ON RETURN:

     W   - REAL(N)
                 THE SINGULAR VALUES OF A.

U   - REAL(MDIM,N)
          THE ORTHOGONAL COLUMNS OF THE MATRIX U IN THE
          DECOMPOSITION. IF MATU=.FALSE., U IS USED AS
          A TEMPORARY ARRAY.

V   - REAL(MDIM,N)
          THE ORTHOGONAL MATRIX V IN THE DECOMPOSITION.
          IF MATV=.FALSE., V IS NOT REFERENCED. IT CAN
          BE A DUMMY (REAL) VARIABLE.

IERR   - INTEGER
          AN ERROR COMPLETION CODE. THE NORMAL
          COMPLETION CODE IS ZERO. IF MORE THAN 30
          ITERATIONS ARE REQUIRED TO DETERMINE A
          SINGULAR VALUE, THE ROUTINE WILL TERMINATE
          WITH IERR SET TO THE INDEX OF THE SINGULAR
          VALUE FOR WHICH THE FAILURE OCCURRED. THE
          SINGULAR VALUES IN THE ARRAY W SHOULD BE
          CORRECT FOR INDICES IERR+1, IERR+2,...,N, AS
          WELL AS THE CORRESPONDING COLUMNS OF U AND V.

RV1   - REAL(N)
          IT IS USED AS TEMPORARY STORAGE TO HOLD THE
          OFF-DIAGONAL ELEMENTS IN THE BIDIAGONAL FORM
          OF A.

```
SUBROUTINE LLSNE(MDIM,M,N,A,B,X,RES,IND)
```

THIS SUBROUTINE COMPUTES THE LEAST-SQUARES SOLUTION OF AN OVERDETERMINED SYSTEM OF LINEAR EQUATIONS A*X = B, WHERE A IS AN M X N (M >= N) MATRIX. THAT IS, IT WILL FIND A VECTOR X WHICH MINIMIZES THE EUCLIDEAN (L2) NORM OF THE RESIDUAL VECTOR R = B - A*X.

THE METHOD USED IS TO FORM THE NORMAL EQUATIONS TRANS(A)*A*X = TRANS(A)*B, AND SOLVE THEM USING THE SUB-ROUTINE PAIR DSPD-SSPD.

CALLING SEQUENCE: CALL LLSNE(MDIM,M,N,A,B,X,RES,IND)

PARAMETERS:

ON ENTRY:

    MDIM   - INTEGER
              THE DECLARED ROW DIMENSION OF THE ARRAY A IN
              THE CALLING PROGRAM. MDIM MUST BE => M.

    M   - INTEGER
              THE NUMBER OF ROWS OF A.

    N   - INTEGER
              THE NUMBER OF COLUMNS OF A. N MUST BE <= 20.

    A   - REAL(MDIM,N)
              THE COEFFICIENT MATRIX OF THE SYSTEM.

    B   - REAL(M)
              THE RIGHT HAND SIDE OF THE SYSTEM.

ON RETURN:

    X   - REAL(N)
              THE LEAST-SQUARES SOLUTION OF THE SYSTEM.

    RES   - REAL(M)
              THE RESIDUAL VECTOR R = B - A*X.

    IND   - INTEGER
              AN INDICATOR OF THE ERROR STATUS.
                 = 0 - NORMAL RETURN,
                 = K - THE LEADING MINOR OF TRANS(A)*A OF
                       ORDER K IS NOT POSITIVE DEFINITE.
                       THE COMPUTATION WAS NOT COMPLETED.

SAMPLE CALLING PROGRAM:

```
      DIMENSION A(30,10),B(30),X(10),RES(30)
         .
         .  -- INPUT OR GENERATE ELEMENTS OF A AND B --
         .
      CALL LLSNE(30,M,N,A,B,X,RES,IND)
         IF (IND .NE. 0) STOP
      DO 10 J = 1,N
         WRITE (6,*) J, X(J)
   10 CONTINUE
      RMSE = SQRT(DPROD(M,RES,RES)/M)
      WRITE (6,*) 'ROOT MEAN SQUARE ERROR =', RMSE
      STOP
      END
```

SUBROUTINE LLSMG(MDIM,M,N,A,B,X,RES,W)

THIS SUBROUTINE COMPUTES THE LEAST SQUARES SOLUTION OF AN OVERDETERMINED LINEAR SYSTEM A*X = B, WHERE A IS AN M X N (M >= N) MATRIX AND B IS AN M-VECTOR. THE METHOD USED IS BASED ON THE MODIFIED GRAM-SCHMIDT ORTHOGONALIZATION PROCESS WHICH FACTORIZES A = Q*R, WHERE Q IS M X N WITH TRANS(Q)*Q = D, AN N X N DIAGONAL MATRIX, AND R IS AN N X N UNIT UPPER TRIANGULAR MATRIX.

THIS SUBROUTINE ASSUMES THAT A IS OF FULL RANK (=N). IF THIS IS NOT TRUE, THEN EXPONENT OVERFLOW MAY OCCUR WITHIN THE ROUTINE AS A RESULT OF DIVISION BY ZERO OR A VERY SMALL NUMBER. THE REMEDY IS TO USE THE SUBROUTINE MGS RATHER THEN LLSMG.

CALLING SEQUENCE: CALL LLSMG(MDIM,M,N,A,B,X,RES,W)

PARAMETERS:

ON ENTRY:

    MDIM   - INTEGER
                THE DECLARED ROW DIMENSION OF THE ARRAY A IN
                THE CALLING PROGRAM. MDIM MUST BE >= M.

    M   - INTEGER
                THE NUMBER OF ROWS OF A.

    N   - INTEGER
                THE NUMBER OF COLUMNS OF A.

    A   - REAL(MDIM,N)
                THE COEFFICIENT MATRIX OF THE SYSTEM.

    B   - REAL(M)
                THE RIGHT HAND SIDE OF THE SYSTEM.

    W   - REAL(N*(N+1)/2)
                WORKSPACE. THIS ARRAY NEED NOT BE INITIALIZED.

ON RETURN:

    A   - THE MATRIX Q OF THE DECOMPOSITION A = Q*R, WHERE
                Q IS ORTHOGONAL AND R IS UNIT UPPER TRIANGULAR.

    X   - REAL(N)
                THE LEAST-SQUARES SOLUTION OF THE SYSTEM.

    RES   - REAL(M)
                THE RESIDUAL VECTOR R = B - A*X.

SAMPLE CALLING PROGRAM:

```
      DIMENSION A(30,10),B(30),X(10),RES(30),W(55)
        .
        .  -- INPUT OR GENERATE ELEMENTS OF A AND B --
        .
      CALL LLSMG(30,M,N,A,B,X,RES,W)
      DO 10 J = 1,N
         WRITE (6,*) J, X(J)
   10 CONTINUE
      RMSE = SQRT(DPROD(M,RES,RES)/M)
      WRITE (6,*) 'ROOT MEAN SQUARE ERROR =', RMSE
      STOP
      END
```

```
      SUBROUTINE MGS(MDIM,M,N,A,B,X,RES,DR,IND)
```

THIS SUBROUTINE COMPUTES THE LEAST SQUARES SOLUTION OF AN OVERDETERMINED LINEAR SYSTEM A*X = B, WHERE A IS AN M X N (M>=N) MATRIX AND B IS AN M-VECTOR. THE METHOD USED IS BASED ON THE MODIFIED GRAM-SCHMIDT ALGORITHM WHICH COMPUTES THE DECOMPOSITION A = Q*R, WHERE Q IS M X N WITH TRANS(Q)*Q = D, AN N X N DIAGONAL MATRIX, AND R IS AN N X N UNIT UPPER TRIANGULAR MATRIX. THE PROCESS CONSISTS OF ORTHOGONALIZING THE COLUMNS OF A SUCCESSIVELY TO FORM Q.

THE ROUTINE INCLUDES A TEST FOR (NUMERICAL) RANK DEFICIENCY, I.E., RANK(A) < N. THE K-TH COLUMN OF A IS DEFINED TO BE DEPENDENT ON THE FIRST K-1 COLUMNS (A IS RANK DEFICIENT) IF

        D(K) + DMAX = DMAX,

WHERE D(K) IS THE K-TH DIAGONAL ENTRY OF THE MATRIX D AND DMAX IS THE MAXIMUM OF D(1), ..., D(K-1). WHEN THIS IS FOUND TO BE TRUE THE ROUTINE WILL EXIT INDICATING WHICH COLUMN IT FOUND TO BE DEPENDENT. IF THE USER WISHES, THE SUBROUTINE CAN BE REENTERED TO CONTINUE THE ORTHOGONALIZATION PROCESS. ALL COLUMNS PREVIOUSLY FOUND TO BE DEPENDENT WILL BE IGNORED AND THE CORRESPONDING ENTRIES IN THE SOLUTION VECTOR X WILL BE SET TO ZERO.

A FEATURE OF THE ROUTINE IS THAT THE USER CAN SPECIFY AT WHICH COLUMN THE ORTHOGONALIZATION IS TO BEGIN. IF, SAY, COLUMN S IS SPECIFIED, THE ROUTINE WILL ASSUME THAT THE PREVIOUS S-1 COLUMNS HAVE ALREADY BEEN ORTHOGONALIZED AND WILL CONTINUE, STARTING WITH COLUMN S. (SEE DESCRIPTION OF THE PARAMETER IND BELOW). THIS FEATURE IS USEFUL IF ADDITIONAL COLUMNS ARE ADDED TO A AFTER A PREVIOUS CALL TO MGS.

CALLING SEQUENCE: CALL MGS(MDIM,M,N,A,B,X,RES,DR,IND)

PARAMETERS:

ON ENTRY:

    MDIM   - INTEGER
                THE DECLARED ROW DIMENSION OF THE ARRAY A IN
                THE CALLING PROGRAM. MDIM MUST BE >= M.

    M   - INTEGER
                THE NUMBER OF ROWS OF A.

    N   - INTEGER
                THE NUMBER OF COLUMNS OF A.

    A   - REAL(MDIM,N)

THE COEFFICENT MATRIX OF THE SYSTEM.

B   - REAL(M)
THE RIGHT HAND SIDE OF THE SYSTEM.

RES   - REAL(M)

DR   - REAL(N*(N+1)/2)
RES AND DR ONLY NEED TO BE INITIALIZED WHEN IND ^= 1, I.E., IF THERE HAS BEEN A PREVIOUS CALL TO MGS. IN THIS CASE, THE INPUT VALUES FOR RES AND DR SHOULD BE THE VALUES THAT WERE RETURNED BY THE PREVIOUS CALL. (IF IND=0, RES DOES NOT HAVE TO BE INITIALIZED.)

IND   - INTEGER
AN INDICATOR OF WHERE THE ORTHOGONALIZATION IS TO BEGIN. THE VALUE OF IND SHOULD BE AS FOLLOWS.
= 1 - IF THIS IS THE FIRST CALL TO MGS.
= S - IF THE FIRST S-1 COLUMNS HAVE ALREADY BEEN ORTHOGONALIZED BY A PREVIOUS CALL TO MGS.
= 0 - IF A HAS ALREADY BEEN DECOMPOSED AND ONLY THE RIGHT-HAND SIDE VECTOR B IS CHANGED.

ON RETURN:

A   - THE MATRIX Q OF THE DECOMPOSITION A = Q*R.

X   - REAL(N)
THE LEAST-SQUARES SOLUTION OF THE SYSTEM. IF THE K-TH COLUMN OF A WAS FOUND TO BE DEPENDENT, THEN X(K) HAS BEEN SET TO 0.

RES   - THE RESIDUAL VECTOR B - A*X.

DR   - THE UPPER TRIANGULAR MATRIX FORMED BY THE PRODUCT D*R WHERE R IS THE UNIT UPPER TRIANGULAR MATRIX IN THE DECOMPOSITION A= Q*R AND D IS THE DIAGONAL MATRIX D = TRANS(Q)*Q.

    DR(I*(I+1)/2) = D(I),
    DR(J*(J-1)/2+I) = D(I)*R(I,J)

IND   - AN INDICATOR OF THE NUMBER OF COLUMNS OF Q AND COMPONENTS OF X THAT WERE COMPUTED.

= N+1, ALL COLUMNS OF A ARE INDEPENDENT AND THE LEAST-SQUARES SOLUTION HAS BEEN SUCCESSFULLY COMPUTED.

= K, 2 <= K <= N. THE FIRST K COLUMNS OF Q AND COMPONENTS OF X HAVE BEEN COMPUTED.

THE ROUTINE HAS EXITED BECAUSE THE K-TH COLUMN OF A WAS FOUND TO BE DEPENDENT. THE VALUE OF THE K-TH DIAGONAL ELEMENT OF D*R HAS BEEN SET TO 0 SO THAT, IN ANY SUBSEQUENT CALL TO MGS, THIS COLUMN WILL BE RECOGNIZED AS BEING DEPENDENT AND THEREFORE OMITTED IN THE COMPUTATIONS.

SAMPLE CALLING PROGRAM:

```
      DIMENSION A(30,10),B(30),X(10),RES(30),DR(55)
         .
         .  -- INPUT OR GENERATE ELEMENTS OF A AND B --
         .
      IND = 1
    5 CALL MGS(30,M,N,A,B,X,RES,DR,IND)
      IF (IND .EQ. N+1) GO TO 10
          WRITE (6,*) IND,'-TH COLUMN IS DEPENDENT'
          IND = IND + 1
          IF (IND .LE. N) GO TO 5
   10 DO 20 J = 1,N
          WRITE (6,*) J, X(J)
   20 CONTINUE
      RMSE = SQRT(DPROD(M,RES,RES)/M)
      WRITE (6,*) 'ROOT MEAN SQUARE ERROR=', RMSE
      STOP
      END
```

```
      SUBROUTINE AMAT(IADIM,IBDIM,ICDIM,M,N,A,B,C)
```

THIS SUBROUTINE COMPUTES THE MATRIX SUM C = A+B, WHERE A, B, AND C ARE M X N MATRICES.

CALLING SEQUENCE: CALL AMAT(IADIM,IBDIM,ICDIM,M,N,A,B,C)

PARAMETERS:

ON ENTRY:

    IADIM,   - INTEGER
     IBDIM,   THE DECLARED ROW DIMENSIONS OF THE ARRAYS
      ICDIM   A, B, AND C, RESPECTIVELY, IN THE CALLING
               PROGRAM.

    M,N   - INTEGER
            THE ROW AND COLUMN DIMENSIONS, RESPECTIVELY,
            OF A, B, AND C.

    A   - REAL(IADIM,N)

    B   - REAL(IBDIM,N)

ON RETURN:

    C   - REAL(ICDIM,N)
            THE MATRIX SUM C = A + B.

```
      SUBROUTINE MMAT(IADIM,IBDIM,ICDIM,M,K,N,A,B,C)
```

THIS SUBROUTINE COMPUTES THE MATRIX PRODUCT C = A*B, WHERE
A IS AN M X K MATRIX, B IS K X N, AND C IS M X N.

CALLING SEQUENCE: CALL MMAT(IADIM,IBDIM,ICDIM,M,K,N,A,B,C)

PARAMETERS:

ON ENTRY:

  IADIM, - INTEGER
  IBDIM, THE DECLARED ROW DIMENSIONS OF THE ARRAYS
  ICDIM   A, B, AND C, RESPECTIVELY, IN THE CALLING
       PROGRAM.

  M,K,N - INTEGER
       AS DESCRIBED ABOVE.

  A   - REAL(IADIM,K)

  B   - REAL(IBDIM,N)

ON RETURN:

  C   - REAL(ICDIM,N)
      THE MATRIX PRODUCT C = A*B.

[Handwritten annotations:]
$A: (a_{ij}) : m \times k$
$B: (b_{ij}) : k \times n$
$C: (c_{ij}) = m \times n$

SUBROUTINE TRANS(MDIM,NDIM,N,M,A,AT)

THIS SUBROUTINE FORMS THE TRANSPOSE OF THE M X N MATRIX A, AND RETURNS IT IN THE N X M ARRAY AT.

CALLING SEQUENCE: CALL TRANS(MDIM,NDIM,M,N,A,AT)

PARAMETERS:

ON ENTRY:

    MDIM,   - INTEGER
     NDIM   THE DECLARED ROW DIMENSIONS OF THE ARRAYS A
              AND AT, RESPECTIVELY, IN THE CALLING PROGRAM.

    M,N   - INTEGER
            THE ROW AND COLUMN DIMENSIONS, RESPECTIVELY,
            OF THE MATRIX A.

    A   - REAL(MDIM,N)
            THE M X N MATRIX TO BE TRANSPOSED.

ON RETURN:

    AT   - REAL(NDIM,M)
            THE TRANSPOSE OF THE MATRIX A.

```
      SUBROUTINE ATRA(MDIM,M,N,A,ATAP)
```

THIS SUBROUTINE COMPUTES THE MATRIX PRODUCT   TRANSPOSE(A)*A,
WHERE A IS M X N, AND RETURNS IT IN PACKED FORM IN THE
ARRAY ATAP. THE ROUTINE UPKS MAY BE USED TO UNPACK ATAP.

CALLING SEQUENCE: CALL ATRA(MDIM,M,N,A,ATAP)

PARAMETERS:

ON ENTRY:

    MDIM   - INTEGER
               THE DECLARED ROW DIMENSION OF THE ARRAY A IN
               THE CALLING PROGRAM.

    M,N   - INTEGER
               THE ROW AND COLUMN DIMENSIONS, RESPECTIVELY,
               OF THE MATRIX A.

    A   - REAL(MDIM,N)
               THE M X N MATRIX FOR WHICH THE   PRODUCT
               TRANSPOSE(A)*A IS SOUGHT.

ON RETURN:

    ATAP   - REAL(N*(N+1)/2)
               THE SYMMETRIC   MATRIX   TRANSPOSE(A)*A,   IN
               PACKED FORM.

        FUNCTION DPROD(N,X,Y)

THE PURPOSE OF THIS FUNCTION SUBPROGRAM IS TO CALCULATE   THE
INNER (ALSO CALLED DOT OR SCALAR) PRODUCT OF TWO VECTORS
OF LENGTH N.   THE DOT PRODUCT IS ACCUMULATED IN DOUBLE
PRECISION   TO   ENSURE   MAXIMUM ACCURACY.   THE DOT PRODUCT IS
DEFINED TO BE ZERO IF N = 0.

CALLING SEQUENCE: DOT = DPROD(N,X,Y)

PARAMETERS:

ON ENTRY:

        N   - INTEGER
                  THE LENGTH OF EACH VECTOR.

        X,Y   - REAL(N)
                  VECTORS WHOSE DOT PRODUCT IS SOUGHT.

ON RETURN:

        DPROD   - REAL
                  THE VALUE OF THE DOT PRODUCT OF X AND Y.

    FUNCTION AMXNM(N,X)

THIS FUNCTION SUBPROGRAM COMPUTES THE L-INFINITY, OR
MAXIMUM, NORM OF A VECTOR X OF LENGTH N. THAT IS, IT
RETURNS

    MAX |X(I)|,   1 <= I <= N
     I

CALLING SEQUENCE: ALINF = AMXNM(N,X)

PARAMETERS:

ON ENTRY:

    N   - INTEGER
                 THE LENGTH OF VECTOR X.

    X   - REAL(N)
                 THE VECTOR FOR WHICH THE L-INFINITY NORM
                 IS SOUGHT.

ON RETURN:

    AMXNM   - REAL
                 THE L-INFINITY NORM OF X.

SUBROUTINE PKS(NDIM,N,A,AP)

THIS SUBROUTINE CONVERTS THE N X N SYMMETRIC MATRIX A INTO PACKED FORM. THE VECTOR AP, CONTAINING THE PACKED FORM OF THE MATRIX, IS ARRANGED AS FOLLOWS:

A(1,1), A(1,2), A(2,2), . . . , A(1,N),..., A(N,N)

CALLING SEQUENCE: CALL PKS(NDIM N,A,AP)

PARAMETERS:

ON ENTRY:

- NDIM — INTEGER
  THE DECLARED ROW DIMENSION OF THE ARRAY A IN THE CALLING PROGRAM.

- N — INTEGER
  THE ORDER OF THE MATRIX A.

- A — REAL(NDIM,N)
  THE N X N SYMMETRIC MATRIX TO BE PUT INTO PACKED FORM.

ON RETURN:

- AP — REAL(N*(N+1)/2)
  VECTOR CONTAINING THE PACKED FORM OF A.

        SUBROUTINE UPKS(AP,NDIM,N,A)

THIS SUBROUTINE IS THE INVERSE OF PKS; THAT IS, IT CONVERTS
AN N X N SYMMETRIC MATRIX FROM PACKED TO FULL STORAGE FORM.

CALLING SEQUENCE: CALL UPKS(AP,NDIM,N,A)

PARAMETERS:

ON ENTRY:

        NDIM   - INTEGER
                    THE DECLARED ROW DIMENSION OF THE ARRAY A   IN
                    THE CALLING PROGRAM.

        N   - INTEGER
                    THE ORDER OF THE MATRIX A.

        AP   - REAL(N*(N+1)/2)
                    VECTOR CONTAINING THE MATRIX A IN PACKED FORM.

ON RETURN:

        A   - REAL(NDIM,N)
                    THE FULL N X N SYMMETRIC MATRIX A.

SUBROUTINE PKSB(NDIM,N,M,A,NPDIM,AP)

THIS SUBROUTINE CONVERTS THE N X N SYMMETRIC BAND MATRIX A
INTO PACKED FORM. THE DIAGONALS OF A BECOME THE ROWS OF THE
PACKED MATRIX AP, THAT IS, AP IS OF THE FORM:

```
    0.0         0.0         0.0    . .   A(1,M+1)  . . .    A(N-M,N)
     .           .           .   .                             .
     .           .           .  .                              .
    0.0         0.0        A(1,3)      .         .     .       .
    0.0        A(1,2)      A(2,3)      .         .     .     A(N-1,N)
   A(1,1)      A(2,2)      A(3,3)      .         .     .      A(N,N)
```

CALLING SEQUENCE: CALL PKSB(NDIM,N,M,A,NPDIM,AP)

PARAMETERS:

ON ENTRY:

    NDIM,   - INTEGER
     NPDIM   THE DECLARED ROW DIMENSIONS OF THE ARRAYS A
              AND AP, RESPECTIVELY, IN THE CALLING PROGRAM.

    N   - INTEGER
            THE ORDER OF THE MATRIX A.

    M   - INTEGER
            THE NUMBER OF DIAGONALS ABOVE THE MAIN
            DIAGONAL OF A (THE BAND WIDTH OF A IS 2*M+1).

    A   - REAL(NDIM,N)
            THE N X N SYMMETRIC BAND MATRIX TO BE PUT
            INTO PACKED FORM.

ON RETURN:

    AP   - REAL(NPDIM,N)
            THE (M+1) X N ARRAY CONTAINING THE PACKED
            FORM OF A.

        SUBROUTINE UPKSB(NPDIM,AP,NDIM,N,M,A)

THIS SUBROUTINE IS THE INVERSE OF PKSB; THAT IS, IT
CONVERTS AN N X N SYMMETRIC BAND MATRIX A FROM PACKED TO
FULL STORAGE FORM.

CALLING SEQUENCE: CALL UPKSB(NPDIM,AP,NDIM,N,M,A)

PARAMETERS:

ON ENTRY:

        NDIM,   - INTEGER
         NPDIM   THE DECLARED ROW DIMENSIONS OF THE ARRAYS A
                    AND AP IN THE CALLING PROGRAM.

        AP   - REAL(NPDIM,N)
                   THE (M+1)*N ARRAY CONTAINING THE PACKED
                   FORM OF A.

        N   - INTEGER
                   THE ORDER OF THE MATRIX A.

        M   - INTEGER
                   THE NUMBER OF DIAGONALS ABOVE THE MAIN
                   DIAGONAL OF A (THE BAND WIDTH OF A IS 2*M+1).

ON RETURN:

        A   - REAL(NDIM,N)
                   THE FULL N X N SYMMETRIC BAND MATRIX A.

```
      SUBROUTINE PITPN(N,X,Y,C)
```

THIS SUBROUTINE COMPUTES THE COEFFICIENTS FOR THE NEWTON FORM OF THE POLYNOMIAL PN(X), OF DEGREE N-1, THAT INTERPOLATES THE DATA (X(I),Y(I)), I = 1,...,N.

```
       PN(X) = C(1)*(X-X(2))*(X-X(3))*...*(X-X(N))
                  +C(2)*(X-X(3))*...*(X-X(N))
                  +...+C(N) .
```

RESTRICTION: SOME OR ALL OF THE INTERPOLATION POINTS MAY COINCIDE BUT THE FOLLOWING CONDITIONS MUST BE SATISFIED -

   IF X(I)=X(I+K), THEN, FOR J=0,1,...,K,

   1. X(I+J) = X(I), AND

   2. Y(I+J) = VALUE OF THE J-TH DERIVATIVE
               OF Y AT X(I).

THE FUNCTION SUBPROGRAM EVNWT CAN BE USED TO EVALUATE PN(X) FOR ANY PRESCRIBED VALUE OF X.

CALLING SEQUENCE: CALL PITPN(N,X,Y,C)

PARAMETERS:

ON ENTRY:

   N   - INTEGER
             THE NUMBER OF DATA POINTS.

   X   - REAL(N)
             THE X-COORDINATES OF THE DATA POINTS.

   Y   - REAL(N)
             THE Y-COORDINATES OF THE DATA POINTS.

ON RETURN:

   C   - REAL(N)
             THE COEFFICIENTS OF THE NEWTON FORM OF THE
             INTERPOLATING POLYNOMIAL.

SAMPLE CALLING PROGRAM:

   SEE THE FUNCTION SUBPROGRAM EVNWT.

```
SUBROUTINE PWLIN(N,X,Y,C)
```

THIS SUBROUTINE COMPUTES THE COEFFICIENTS OF THE CONTINUOUS, PIECEWISE LINEAR POLYNOMIAL G1(X) THAT INTERPOLATES THE DATA (X(I),Y(I)), I = 1,...,N.

THE FUNCTION SUBPROGRAM EPLIN CAN BE USED TO EVALUATE G1(X).

CALLING SEQUENCE: CALL PWLIN(N,X,Y,C)

PARAMETERS:

ON ENTRY:

    N   - INTEGER
             THE NUMBER OF DATA POINTS.

    X   - REAL(N)
             THE X-COORDINATES OF THE DATA POINTS. THE POINTS NEED NOT BE EQUALLY SPACED, BUT MUST BE ORDERED: X(1) < X(2) < ... < X(N).

    Y   - REAL(N)
             THE Y-COORDINATES OF THE DATA POINTS.

ON RETURN:

    C   - REAL(2,N)
             THE COEFFICIENTS OF G1(X). THE J-TH COLUMN OF C CONTAINS THE J-TH PIECE, P1J(X), 1 <= J <= N-1. THAT IS, IF X(J) <= X <= X(J+1), THEN

             G1(X) = P1J(X) = C(1,J) + C(2,J)*(X-X(J)).

SAMPLE CALLING PROGRAM:

    SEE THE FUNCTION SUBPROGRAM EPLIN.

        SUBROUTINE PHERM(N,X,Y,YP,C)

THIS SUBROUTINE COMPUTES THE COEFFICIENTS OF THE PIECEWISE
CUBIC HERMITE POLYNOMIAL G3(X) THAT INTERPOLATES THE DATA
(X(I), Y(I), Y'(I)), I = 1,...,N.

THE FUNCTION SUBPROGRAM EPCUB CAN BE USED TO EVALUATE G3(X).

CALLING SEQUENCE: CALL PHERM(N,X,Y,YP,C)

PARAMETERS:

ON ENTRY:

        N   - INTEGER
                   THE NUMBER OF DATA POINTS.

        X   - REAL(N)
                   THE X-COORDINATES OF THE DATA POINTS. THE
                   POINTS NEED NOT BE EQUALLY SPACED, BUT MUST
                   BE ORDERED: X(1) < X(2) < ... < X(N) .

        Y   - REAL(N)
                   THE Y-COORDINATES OF THE DATA POINTS.

        YP   - REAL(N)
                   THE VALUES OF THE DERIVATIVE AT THE DATA PTS.

ON RETURN:

        C   - REAL(4,N)
                   THE COEFFICIENTS OF G3(X). THE J-TH COLUMN
                   OF C CONTAINS THE J-TH PIECE, P3J(X),
                   1 <= J <= N-1. THAT IS, IF X(J) <= X <=
                   X(J+1), THEN

                     G3(X) = P3J(X) = C(1,J) + C(2,J)*ZJ
                                    + C(3,J)*ZJ**2 + C(4,J)*ZJ**3,

                   WHERE ZJ = (X-X(J)).

SAMPLE CALLING PROGRAM:

    SEE THE FUNCTION SUBPROGRAM EPCUB.

```
SUBROUTINE SPLN(N,X,Y,C)
```

THIS SUBROUTINE COMPUTES THE COEFFICIENTS FOR THE CUBIC SPLINE S3(X) THAT INTERPOLATES THE DATA (X(I), Y(I)), I = 1,...,N. UP TO 100 DATA POINTS MAY BE USED.

A CUBIC SPLINE INTERPOLATION ROUTINE NEEDS AN APPROXIMATION TO THE DERIVATIVE AT EACH END POINT X(1) AND X(N). IN SPLN, THESE ARE OBTAINED BY QUADRATIC INTERPOLATION AT THE FIRST 3 AND LAST 3 DATA POINTS AND THEN EVALUATING THE DERIVATIVE OF EACH APPROXIMATION AT X(1) AND X(N), RESPECTIVELY. IF THE DERIVATIVE VALUES ARE ALREADY KNOWN, THEN THE OTHER CUBIC SPLINE INTERPOLATION ROUTINE - SPLNF - SHOULD BE USED.

THE FUNCTION SUBPROGRAM EPCUB CAN BE USED TO EVALUATE G3(X).

CALLING SEQUENCE: CALL SPLN(N,X,Y,C)

PARAMETERS:

ON ENTRY:

    N   - INTEGER
              THE NUMBER OF DATA POINTS. N MUST BE <= 100.

    X   - REAL(N)
              THE X-COORDINATES OF THE DATA POINTS. THE POINTS NEED NOT BE EQUALLY SPACED, BUT MUST BE ORDERED: X(1) < X(2) < ... < X(N) .

    Y   - REAL(N)
              THE Y-COORDINATES OF THE DATA POINTS.

ON RETURN:

    C   - REAL(4,N)
              THE COEFFICIENTS OF S3(X). THE J-TH COLUMN OF C CONTAINS THE J-TH PIECE, P3J(X), 1 <= J <= N-1. THAT IS, IF X(J) <= X <= X(J+1), THEN

$$S3(X) = P3J(X) = C(1,J) + C(2,J)*ZJ + C(3,J)*ZJ**2 + C(4,J)*ZJ**3,$$

              WHERE ZJ = (X-X(J)).

SAMPLE CALLING PROGRAM:

    SEE THE FUNCTION SUBPROGRAM EPCUB.

```
      SUBROUTINE SPLNF(N,X,Y,YP1,YPN,C)
```

THIS SUBROUTINE COMPUTES THE COEFFICIENTS FOR THE CUBIC SPLINE S3(X) THAT INTERPOLATES THE DATA (X(I), Y(I)), I = 1,...,N. UP TO 100 DATA POINTS MAY BE USED.

THIS PARTICULAR VERSION OF CUBIC SPLINE INTERPOLATION ASSUMES THAT THE VALUES OF Y'(X(1)) AND Y'(X(N)), OR SOME APPROXIMATIONS TO THEM, ARE GIVEN. IF NOT, THEN THE OTHER SPLINE INTERPOLATION ROUTINE - SPLN - SHOULD BE USED.

THE FUNCTION SUBPROGRAM EPCUB CAN BE USED TO EVALUATE G3(X).

CALLING SEQUENCE: CALL SPLNF(N,X,Y,YP1,YPN,C)

PARAMETERS:

ON ENTRY:

- N   - INTEGER
             THE NUMBER OF DATA POINTS. N MUST BE <=100.

- X   - REAL(N)
             THE X-COORDINATES OF THE DATA POINTS. THE POINTS NEED NOT BE EQUALLY SPACED, BUT MUST BE ORDERED: X(1) < X(2) < ... < X(N) .

- Y   - REAL(N)
             THE Y-COORDINATES OF THE DATA POINTS.

- YP1, YPN - REAL
             THE VALUES OF THE DERIVATIVE AT X(1) AND X(N) RESPECTIVELY.

ON RETURN:

- C   - REAL(4,N)
             THE COEFFICIENTS OF S3(X). THE J-TH COLUMN OF C CONTAINS THE J-TH PIECE, P3J(X), 1 <= J <= N-1. THAT IS, IF X(J) <= X <= X(J+1), THEN

```
             S3(X) = P3J(X) = C(1,J) + C(2,J)*ZJ
                            + C(3,J)*ZJ**2 + C(4,J)*ZJ**3,
```

             WHERE ZJ = (X-X(J)).

SAMPLE CALLING PROGRAM:

    SEE THE FUNCTION SUBPROGRAM EPCUB.

```
      SUBROUTINE LSQCS(M,X,Y,N,XKNOT,C,RES)
```

THIS SUBROUTINE COMPUTES THE BEST LEAST SQUARES CUBIC SPLINE
APPROXIMATION S3(X) FOR THE DATA (X(I), Y(I)), 1 <= I <= M.
THE SPLINE HAS N KNOTS WHICH ARE FIXED AT THE POINTS
XKNOT(J), 1 <= J <= N. THE KNOTS MUST BE DISTINCT AND FORM
A BRACKET FOR THE DATA, THAT IS,

    XKNOT(1) <= X(I) <= XKNOT(N), 1 <= I <= M.

THE FUNCTION SUBPROGRAM EPCUB CAN BE USED TO EVALUATE S3(X).

CALLING SEQUENCE: CALL LSQCS(M,X,Y,N,XKNOT,C,RES)

PARAMETERS:

ON ENTRY:

    M   - INTEGER
                THE NUMBER OF DATA POINTS. M MUST BE <= 100

    X,Y   - REAL(M)
                THE X- AND Y-COORDINATES, RESPECTIVELY, OF
                THE DATA POINTS.

    N   - INTEGER
                THE NUMBER OF KNOTS. N MUST BE <= 20

    XKNOT   - REAL(N)
                THE X-COORDINATES OF THE KNOTS. THEY MUST BE
                ORDERED: XKNOT(1) < XKNOT(2) <...< XKNOT(N).

ON RETURN:

    C   - REAL(4,N)
                THE COEFFICIENTS OF S3(X). THE J-TH COLUMN
                OF C CONTAINS THE J-TH PIECE, P3J(X),
                1 <= J <= N-1. THAT IS, IF XKNOT(J) <=
                X <= XKNOT(J+1), THEN

                    S3(X) = P3J(X) = C(1,J) + C(2,J)*ZJ
                                   + C(3,J)*ZJ**2 + C(4,J)*ZJ**3,

                WHERE ZJ = (X-XKNOT(J)).

    RES   - REAL(M)
                THE RESIDUALS, THAT IS

                    RES(X(I)) = Y(I) - S3(X(I)), 1 <= I <= M.

SAMPLE CALLING PROGRAM:

```
      DIMENSION X(50),Y(50),XKNOT(5),C(4,5),RES(50)
      READ (5,*) M, (X(I), Y(I), I=1,M)
      READ (5,*) N, (XKNOT(J), J=1,N)
C
C     *************************
C     *
C     *  COMPUTE S3(X) AND DETERMINE THE ROOT MEAN
C     *  SQUARE ERROR.
C     *
C     *************************
C
      CALL LSQCS(M,X,Y,N,XKNOT,C,RES)
      NM1 = N - 1
      DO 10 J = 1, NM1
         WRITE (6,*) J,XKNOT(J),(C(I,J),I=1,4)
   10 CONTINUE
      RMSE = SQRT(DPROD(M,RES,RES)/M)
      WRITE (6,*) 'ROOT MEAN SQUARE ERROR = ', RMSE
C
C     *************************
C     *
C     *  EVALUATE S3(X) AT X = XBAR
C     *
C     *************************
C
      YBAR = EPCUB(N,XBAR,XKNOT,C)
       .
       .
      STOP
      END
```

```
      FUNCTION EVNWT(N,XBAR,X,C)
```

THIS ROUTINE WILL EVALUATE THE NEWTON FORM OF A POLYNOMIAL
PN(X) AT A SPECIFIED VALUE X = XBAR; THAT IS, IT COMPUTES

```
    PN(XBAR) =
      C(1)*(XBAR-X(2))*(XBAR-X(3))*...*(XBAR-X(N))
               +C(2)*(XBAR-X(3))*...*(XBAR-X(N))
               + ...   +C(N)
```

THE COEFFICIENTS C(I) CAN BE GENERATED BY INTERPOLATION
USING THE SUBROUTINE PITPN.

CALLING SEQUENCE: YBAR = EVNWT(N,XBAR,X,C)

PARAMETERS:

ON ENTRY:

       N   - INTEGER
              THE ORDER OF PN(X) IS N-1.

       XBAR   - REAL
              THE POINT AT WHICH THE POLYNOMIAL IS TO BE
              EVALUATED.

       X   - REAL(N)
              THE X VALUES USED TO DEFINE THE NEWTON FORM
              OF PN(X).

       C   - REAL(N)
              THE COEFFICIENTS DEFINING PN(X).

ON RETURN:

       EVNWT - REAL
              THE VALUE PN(XBAR).

SAMPLE CALLING PROGRAM:

```
      DIMENSION X(10),Y(10),C(10)
      READ (5,*) N,(X(I),Y(I),I=1,N)
C
C     ************************
C     *
C     *   GENERATE THE INTERPOLATING POLYNOMIAL PN(X).
C     *
C     ************************
C
      CALL PITPN(N,X,Y,C)
C
```

```
C     *************************
C     *
C     *   EVALUATE PN(X) AT X=XBAR.
C     *
C     *************************
C
      YBAR = EVNWT(N,XBAR,X,C)
         .
         .
      STOP
      END
```

          FUNCTION EPLIN(N,XBAR,X,C)

THIS FUNCTION SUBPROGRAM EVALUATES A CONTINUOUS PIECEWISE
LINEAR POLYNOMIAL G1(X) WITH KNOTS AT X(1), X(2), ..., X(N),
AT A SPECIFIED VALUE X = XBAR. THAT IS, IT DETERMINES THE
SUBINTERVAL (X(J),X(J+1)) CONTAINING XBAR AND THEN COMPUTES

        G1(XBAR) = C(1,J) + C(2,J)*(XBAR-X(J)).

A BINARY SEARCH IS USED TO DETERMINE THE CORRECT
SUBINTERVAL. THE COEFFICIENTS C(I,J) CAN BE GENERATED BY
INTERPOLATION USING THE SUBROUTINE PWLIN.

CALLING SEQUENCE: YBAR = EPLIN(N,XBAR,X,C)

PARAMETERS:

ON ENTRY:

        N   - INTEGER
                  THE NUMBER OF KNOTS.

        XBAR   - REAL
                  THE POINT AT WHICH G1(X) IS TO BE EVALUATED.

        X   - REAL(N)
                  THE X-COORDINATES OF THE KNOTS.

        C   - REAL(2,N)
                  THE COEFFICIENTS DEFINING EACH PIECE
                  OF G1(X).

ON RETURN:

        EPLIN   - REAL
                  THE VALUE G1(XBAR). IF XBAR IS TO THE LEFT
                  (RIGHT) OF X(1) (X(N)), THEN THE FIRST (LAST)
                  PIECE IS USED TO DETERMINE G1(XBAR).

SAMPLE CALLING PROGRAM:

      DIMENSION X(50), Y(50),C(2,50)
      READ (5,*) N,(X(I),Y(I),I=1,N)
C
C   *************************
C   *
C   * GENERATE THE INTERPOLATING PIECEWISE LINEAR
C   * POLYNOMIAL G1(X).
C   *
C   *************************
C

```
      CALL  PWLIN(N,X,Y,C)
C
C     *************************
C     *
C     *   EVALUATE G1(X) AT X=XBAR.
C     *
C     *************************
C
      YBAR = EPLIN(N,XBAR,X,C)
         .
         .
      STOP
      END
```

```
      FUNCTION EPCUB(N,XBAR,X,C)
```

THIS FUNCTION SUBPROGRAM EVALUATES A CONTINUOUS PIECEWISE CUBIC POLYNOMIAL G3(X) WITH KNOTS AT X(1), X(2), ..., X(N), AT A SPECIFIED VALUE X = XBAR. THAT IS, IT DETERMINES THE SUBINTERVAL (X(J), X(J+1)) CONTAINING XBAR AND THEN COMPUTES

      G3(XBAR) = C(1,J) + C(2,J)*ZJBAR
               + C(3,J)*ZJBAR**2 + C(4,J)*ZJBAR**3,

WHERE ZJBAR = XBAR - X(J). A BINARY SEARCH IS USED TO DETERMINE THE CORRECT SUBINTERVAL. THE COEFFICIENTS C(I,J) CAN BE GENERATED EITHER BY INTERPOLATION USING ANY ONE OF THE SUBROUTINES PHERM, SPLN OR SPLNF, OR ELSE BY LEAST SQUARES APPROXIMATION USING LSQCS.

CALLING SEQUENCE: YBAR = EPCUB(N,XBAR,X,C)

PARAMETERS:

ON ENTRY:

        N   - INTEGER
                  THE NUMBER OF KNOTS.
        XBAR   - REAL
                  THE POINT AT WHICH G3(X) IS TO BE EVALUATED.

        X   - REAL(N)
                  THE X-COORDINATES OF THE KNOTS.

        C   - REAL(4,N)
                  THE COEFFICIENTS DEFINING EACH PIECE OF G3(X).

ON RETURN:

        EPCUB   - REAL
                  THE VALUE G3(XBAR). IF XBAR IS TO THE LEFT
                  (RIGHT) OF X(1) (X(N)), THEN THE FIRST (LAST)
                  PIECE IS USED TO DETERMINE G3(XBAR).

SAMPLE CALLING PROGRAM:

      DIMENSION X(50),C(4,50),Y(50),YP(50)
         .
         . -- INPUT OR OTHERWISE GENERATE THE DATA --
         .
C
C   *************************
C   *
C   *   GENERATE PIECEWISE CUBIC INTERPOLATING
C   *   FUNCTION G3(X).

```
C     *
C     ************************
C
      CALL PHERM(N,X,Y,YP,C) ... OR SPLN (N,X,Y,C)
                                    SPLNF(N,X,Y,YP1,YPN,C)
                                    LSQCS(M,X,Y,N,XKNOT,C,RES)
C
C     ************************
C     *
C     *   EVALUATE G3(X) AT X=XBAR.
C     *
C     ************************
C
      YBAR=EPCUB(N,XBAR,X,C)
         .
         .
      STOP
      END
```

```
      SUBROUTINE BSECT(F,A,B,XTOL,FTOL,ROOT,IND,LIST)
```

THIS SUBROUTINE USES THE BISECTION METHOD TO FIND A (REAL) SOLUTION OF THE SINGLE NONLINEAR EQUATION F(X) = 0. GIVEN AN INITIAL BRACKET (A,B) ON WHICH THE FUNCTION F(X) CHANGES SIGN, THE ROUTINE COMPUTES A SEQUENCE OF NESTED BRACKETS (AR, BR), OF LENGTH DELTAR = BR - AR, UNTIL ONE OF THE FOLLOWING TERMINATION CONDITIONS IS SATISFIED:

    1. |DELTAR| <= XTOL*(1.0+|XR|),
        WHERE XR IS THE MIDPOINT OF THE CURRENT BRACKET (AR, BR);

    2. |F(XR)| <= FTOL.

THE ROUTINE INCLUDES A FACILITY TO OUTPUT INTERMEDIATE RESULTS. THIS FACILITY IS CONTROLLED THROUGH THE PARAMETER LIST (SEE BELOW).

CALLING SEQUENCE: CALL BSECT(F,A,B,XTOL,FTOL,ROOT,IND,LIST)

PARAMETERS:

ON ENTRY:

    F   - REAL
           A USER-SUPPLIED FUNCTION SUBPROGRAM, F(X), FOR EVALUATING THE FUNCTION F AT ANY GIVEN X IN THE INTERVAL (A,B).

    A,B   - REAL
           THE LEFT AND RIGHT ENDPOINTS, RESPECTIVELY, OF THE INITIAL BRACKET.

    XTOL   - REAL
           THE ACCEPTABLE TOLERANCE IN THE SIZE OF THE FINAL BRACKET.

    FTOL   - REAL
           THE ACCEPTABLE TOLERANCE IN |F(XR)|.

    LIST   - INTEGER
           A PARAMETER USED TO CONTROL OUTPUT OF INTERMEDIATE RESULTS:
           = 0   - NO OUTPUT;
           = K   - OUTPUT EVERY K-TH ITERATION.

ON RETURN:

    A,B   - THE ENDPOINTS OF THE MOST RECENT BRACKET.

    ROOT   - THE MOST RECENT APPROXIMATE SOLUTION, THAT IS,

                    THE MIDPOINT OF THE MOST RECENT BRACKET.

        IND   - INTEGER
                  THE STATUS OF THE RESULT:
                  = 1   TERMINATION CONDITION (1) IS SATISFIED;
                  = 2   TERMINATION CONDITION (2) IS SATISFIED;
                  = 12 BOTH TERMINATION CONDITIONS (1) AND (2)
                        ARE SATISFIED.

        LIST   - THE NUMBER OF ITERATIONS PERFORMED. IF LIST
                  IS -1, THEN NO ITERATIONS WERE PERFORMED
                  BECAUSE THE INITIAL VALUES OF A AND B DID NOT
                  FORM A BRACKET.

SAMPLE CALLING PROGRAM:

    EXTERNAL F
    READ (5,*) A,B
    XTOL = 1.E-5
    FTOL = 1.E-5
    LIST = 0
    CALL   BSECT(F,A,B,XTOL,FTOL,ROOT,IND,LIST)
    WRITE (6,*) ROOT,IND,LIST
    STOP
    END

    FUNCTION F(X)
    F = ... EXPRESSION FOR EVALUATING F
    RETURN
    END

```
SUBROUTINE SECNT(F,XM1,X0,XTOL,FTOL,ITMAX,IND,LIST)
```

THIS SUBROUTINE USES THE SECANT METHOD TO FIND A (REAL) SOLUTION OF THE SINGLE NONLINEAR EQUATION F(X) = 0. GIVEN TWO INITIAL GUESSES XM1, X0 FOR THE SOLUTION, THE ROUTINE COMPUTES SUCCESSIVE ITERATES XR UNTIL ONE OF THE FOLLOWING TERMINATION CONDITIONS IS SATISFIED:

    1. |XR - XRM1| <= XTOL*(1.0+|XR|),
         WHERE XRM1, XR ARE THE TWO MOST RECENT ITERATES;

    2. |F(XR)| <= FTOL;

    3.   ITMAX ITERATIONS HAVE BEEN PERFORMED.

THE ROUTINE INCLUDES A FACILITY TO OUTPUT INTERMEDIATE RESULTS. THIS FACILITY IS CONTROLLED THROUGH THE PARAMETER LIST (SEE BELOW).

CALLING SEQUENCE: CALL SECNT(F,XM1,X0,XTOL,FTOL,ITMAX,IND,LIST)

PARAMETERS:

ON ENTRY:

    F   - REAL
              A USER-SUPPLIED FUNCTION SUBPROGRAM, F(X), FOR EVALUATING THE FUNCTION F AT ANY GIVEN X.

    XM1,   - REAL
     X0   INITIAL GUESSES FOR THE SOLUTION.

    XTOL   - REAL
              THE ACCEPTABLE TOLERANCE IN THE CHANGE |XR - XRM1| BETWEEN SUCCESSIVE ITERATES.

    FTOL   - REAL
              THE ACCEPTABLE TOLERANCE IN |F(XR)|.

    ITMAX   - INTEGER
              THE MAXIMUM NUMBER OF ITERATIONS TO BE PERFORMED.

    LIST   - INTEGER
              A PARAMETER USED TO CONTROL OUTPUT OF INTERMEDIATE RESULTS:
              = 0   - NO OUTPUT;
              = K   - OUTPUT EVERY K-TH ITERATION.

ON RETURN:

    XM1,   - THE TWO MOST RECENT ITERATES. THE APPROX-
     X0

IMATE SOLUTION IS X0 SINCE IT IS THE MOST RECENT ITERATE.

ITMAX   — THE NUMBER OF ITERATIONS PERFORMED.

IND   — INTEGER
          THE STATUS OF THE RESULT:
          = 1   TERMINATION CONDITION (1) IS SATISFIED;
          = 2   TERMINATION CONDITION (2) IS SATISFIED;
          = 12 BOTH TERMINATION CONDITIONS (1) AND (2) ARE SATISFIED;
          = 3   CONVERGENCE COULD NOT BE ACHIEVED WITHIN ITMAX ITERATIONS.
          = 4   THE ITERATION TERMINATED BECAUSE THE SECANT LINE BECAME <u>HORIZONTAL</u> (¦SLOPE¦ < 1.E-30).

SAMPLE CALLING PROGRAM:

```
      EXTERNAL F
      READ (5,*) XM1,X0
      XTOL = 1.E-5
      FTOL = 1.E-5
      ITMAX = 20
      LIST = 0
      CALL SECNT(F,XM1,X0,XTOL,FTOL,ITMAX,IND,LIST)
      WRITE(6,*) XM1,X0,ITMAX,IND
      STOP
      END

      FUNCTION F(X)
      F = ... EXPRESSION FOR EVALUATING F
      RETURN
      END
```

```
SUBROUTINE NWTN(F,FPRIME,X0,XTOL,FTOL,ITMAX,IND,LIST)
```

THIS SUBROUTINE USES THE NEWTON-RAPHSON METHOD TO FIND A (REAL) SOLUTION OF THE SINGLE NONLINEAR EQUATION F(X) = 0. GIVEN AN INITIAL GUESS X0, THE ROUTINE COMPUTES SUCCESSIVE ITERATES XR UNTIL ONE OF THE FOLLOWING TERMINATION CONDITIONS IS SATISFIED:

1. $|XR - XRM1| \leq XTOL*(1.0+|XR|)$,
   WHERE XRM1, XR ARE THE TWO MOST RECENT ITERATES;

2. $|F(XR)| \leq FTOL$;

3. ITMAX ITERATIONS HAVE BEEN PERFORMED.

THE ROUTINE INCLUDES A FACILITY TO OUTPUT INTERMEDIATE RESULTS. THIS FACILITY IS CONTROLLED THROUGH THE PARAMETER LIST (SEE BELOW).

CALLING SEQUENCE: CALL NWTN(F,FPRIME,X0,XTOL,FTOL,ITMAX,
                               IND,LIST)

PARAMETERS:

ON ENTRY:

    F   - REAL
               A USER SUPPLIED FUNCION SUBPROGRAM, F(X),
               FOR EVALUATING THE FUNCTION F AT ANY POINT X.

    FPRIME   - REAL
               A USER SUPPLIED FUNCTION SUBPROGRAM,
               FPRIME(X), FOR EVALUATING THE DERIVATIVE
               F'(X) OF F AT ANY POINT X.

    X0   - REAL
               INITIAL GUESS FOR THE SOLUTION.

    XTOL   - REAL
               THE ACCEPTABLE TOLERANCE IN THE CHANGE
               $|XR - XRM1|$ BETWEEN SUCCESSIVE ITERATES.

    FTOL   - REAL
               THE ACCEPTABLE TOLERANCE IN $|F(XR)|$.

    ITMAX   - INTEGER
               THE MAXIMUM NUMBER OF ITERATIONS TO BE
               PERFORMED.

    LIST   - INTEGER
               A PARAMETER USED TO CONTROL OUTPUT OF
               INTERMEDIATE RESULTS:

```
                        = 0    - NO OUTPUT;
                        = K    - OUTPUT EVERY K-TH ITERATION.
```

ON RETURN:

    XO   - THE MOST RECENT APPROXIMATE SOLUTION.

    ITMAX - THE NUMBER OF ITERATIONS PERFORMED.

    IND   - INTEGER
```
             THE STATUS OF THE RESULT:
             = 1    TERMINATION CONDITION (1) IS SATISFIED;
             = 2    TERMINATION CONDITION (2) IS SATISFIED;
             = 12   BOTH TERMINATION CONDITIONS (1) AND (2)
                    ARE SATISFIED;
             = 3    CONVERGENCE COULD NOT BE ACHIEVED
                    WITHIN ITMAX ITERATIONS;
             = 4    THE ITERATION TERMINATED BECAUSE THE
                    DERIVATIVE VANISHED (<1.E-30).
```

SAMPLE CALLING PROGRAM:

```
     EXTERNAL F,FPRIME
     READ (5,*) XO
     XTOL = 1.E-5
     FTOL = 1.E-5
     ITMAX = 15
     LIST = 0
     CALL NWTN(F,FPRIME,XO,XTOL,FTOL,ITMAX,IND,LIST)
     WRITE(6,*) XO,ITMAX,IND
     STOP
     END

     FUNCTION F(X)
     F = ... EXPRESSION FOR EVALUATING F
     RETURN
     END

     FUNCTION FPRIME(X)
     FPRIME = ... EXPRESSION FOR EVALUATION F'
     RETURN
     END
```

```
      SUBROUTINE MULR(F,X0,XTOL,FTOL,ITMAX,IND,LIST)
```

THIS SUBROUTINE USES MULLER'S METHOD TO FIND A (REAL OR COMPLEX) SOLUTION OF THE SINGLE NONLINEAR EQUATION F(X) = 0. GIVEN AN INITIAL GUESS X0 (WHICH NEED NOT BE VERY ACCURATE) THE ROUTINE AUTOMATICALLY GENERATES TWO MORE GUESSES AND THEN COMPUTES SUCCESSIVE ITERATES XR UNTIL ONE OF THE FOLLOWING TERMINATION CONDITIONS IS SATISFIED:

    1. |XR - XRM1| <= XTOL*(1.0+|XR|),
         WHERE XRM1, XR ARE THE TWO MOST RECENT ITERATES;

    2. |F(XR)| <= FTOL;

    3.   ITMAX ITERATIONS HAVE BEEN PERFORMED.

THE ROUTINE INCLUDES A FACILITY TO OUTPUT INTERMEDIATE RESULTS. THIS FACILITY IS CONTROLLED THROUGH THE PARAMETER LIST (SEE BELOW).

CALLING SEQUENCE: CALL MULR(F,X0,XTOL,FTOL,ITMAX,IND,LIST)

PARAMETERS:

ON ENTRY:

    F   - COMPLEX
          A USER-SUPPLIED FUNCTION SUBPROGRAM, F(X), FOR EVALUATING THE FUNCTION F AT ANY POINT X.

    X0   - COMPLEX
          INITIAL GUESS FOR THE SOLUTION.

    XTOL - REAL
          THE ACCEPTABLE TOLERANCE IN THE CHANGE |XR - XRM1| BETWEEN SUCCESSIVE ITERATES.

    FTOL - REAL
          THE ACCEPTABLE TOLERANCE IN |F(XR)|.

    ITMAX - INTEGER
          THE MAXIMUM NUMBER OF ITERATIONS TO BE PERFORMED.

    LIST - INTEGER
          A PARAMETER USED TO CONTROL OUTPUT OF INTERMEDIATE RESULTS:
          = 0   - NO OUTPUT;
          = K   - OUTPUT EVERY K-TH ITERATION.

ON RETURN:

XO   - THE MOST RECENT APPROXIMATE SOLUTION.

ITMAX   - THE NUMBER OF ITERATIONS PERFORMED.

IND   - INTEGER
            THE STATUS OF THE RESULT:
            = 1   TERMINATION CONDITION (1) IS SATISFIED;
            = 2   TERMINATION CONDITION (2) IS SATISFIED;
            = 12 BOTH TERMINATION CONDITIONS (1) AND (2)
                  ARE SATISFIED;
            = 3   CONVERGENCE COULD NOT BE ACHIEVED
                  WITHIN ITMAX ITERATIONS.

SAMPLE CALLING PROGRAM:

```
      COMPLEX  XO
      REAL XTOL,FTOL
      EXTERNAL F
      READ (5,*) XO
      XTOL = 1.E-5
      FTOL = 1.E-5
      ITMAX = 15
      LIST = 0
      CALL  MULR(F,XO,XTOL,FTOL,ITMAX,IND,LIST)
      WRITE(6,*) XO,ITMAX,IND
      STOP
      END

      COMPLEX FUNCTION F(X)
      F = ... EXPRESSION FOR EVALUATING F
      RETURN
      END
```

```
      SUBROUTINE ZERO(F,A,B,XTOL,FTOL,ROOT,IND,LIST)
```

THIS SUBROUTINE USES BRENT'S ALGORITHM TO FIND A (REAL) SOLUTION OF THE SINGLE NONLINEAR EQUATION F(X) = 0. GIVEN AN INITIAL BRACKET (A,B) ON WHICH THE FUNCTION F(X) CHANGES SIGN, THE ROUTINE SUCCESSIVELY COMPUTES A SEQUENCE OF NESTED BRACKETS (AR,BR), OF LENGTH DELTAR, UNTIL ONE OF THE FOLLOWING TERMINATION CONDITIONS IS SATISFIED:

    1. |DELTAR| <= XTOL*(1.0+|XR|),
         WHERE XR, A POINT IN THE CURRENT BRACKET (AR,BR),
         IS THE MOST RECENT APPROXIMATE SOLUTION;

    2. |F(XR)| <= FTOL.

THIS ROUTINE IS A MODIFIED VERSION OF THE FUNCTION SUBPROGRAM ZEROIN.
(SEE: COMPUTER METHODS FOR MATHEMATICAL COMPUTATIONS,
      FORSYTHE, MALCOLM, AND MOLER,
      PRENTICE-HALL, INC. (1977).   )

CALLING SEQUENCE: CALL ZERO(F,A,B,XTOL,FTOL,ROOT,IND,LIST)

PARAMETERS:

ON ENTRY:

     F   - REAL
             A USER-SUPPLIED FUNCTION SUBPROGRAM, F(X),
             FOR EVALUATING THE FUNCTION F AT ANY GIVEN X
             IN THE INTERVAL (A,B).

     A,B   - REAL
             THE ENDPOINTS OF THE INITIAL BRACKET.

     XTOL   - REAL
             THE ACCEPTABLE TOLERANCE IN THE SIZE OF THE
             FINAL BRACKET.

     FTOL   - REAL
             THE ACCEPTABLE TOLERANCE IN |F(XR)|.

     LIST   - INTEGER
             A PARAMETER USED TO CONTROL THE OUTPUT OF
             INTERMEDIATE RESULTS:
             = 0 - NO OUTPUT;
             = K - OUTPUT EVERY K-TH ITERATION.

             IN THE OUTPUT, THE COLUMN HEADED 'ITYPE'
             INDICATES WHICH METHOD WAS USED:

             ITYPE = 1 - INVERSE QUADRATIC INTERPOLATION;

                            2 - LINEAR INTERPOLATION (SECANT
                                METHOD);
                            3 - BISECTION METHOD.

ON RETURN:

    A,B   - THE ENDPOINTS OF THE MOST RECENT BRACKET.

    ROOT   - THE MOST RECENT APPROXIMATE SOLUTION.

    IND   - INTEGER
              THE STATUS OF THE RESULT:
              = 1   TERMINATION CONDITION (1) IS SATISFIED;
              = 2   TERMINATION CONDITION (2) IS SATISFIED;
              = 12 BOTH TERMINATION CONDITIONS (1) AND (2)
                    ARE SATISFIED.

    LIST   - THE NUMBER OF ITERATIONS PERFORMED. IF LIST
              IS -1, THEN NO ITERATIONS WERE PERFORMED
              BECAUSE THE INITIAL VALUES OF A AND B DID NOT
              FORM A BRACKET.

SAMPLE CALLING PROGRAM:

```
      EXTERNAL F
      READ (5,*) A,B
      XTOL = 1.E-5
      FTOL = 1.E-5
      LIST = 0
      CALL  ZERO(F,A,B,XTOL,FTOL,ROOT,IND,LIST)
      WRITE (6,*) ROOT,IND,LIST
      STOP
      END

      FUNCTION F(X)
      F = ... EXPRESSION FOR EVALUATING F
      RETURN
      END
```

```
      SUBROUTINE LAGR(N,A,XREAL,XIMAG,IND,LIST)
```

THIS SUBROUTINE USES LAGUERRE'S METHOD TO FIND ALL N SOLUTIONS (REAL OR COMPLEX) OF THE SINGLE POLYNOMIAL EQUATION

$$P(X) = A(1)*X^N + A(2)*X^{N-1} + \ldots + A(N)*X + A(N+1) = 0$$

OF DEGREE N (<= 20) WITH REAL COEFFICIENTS.

A DEFLATION PROCESS IS USED IN ORDER TO AVOID REPEATED COMPUTATION OF THE SAME SOLUTION. WHEN A REAL SOLUTION RJ IS FOUND, THE ROUTINE DEFLATES BY THE LINEAR FACTOR (X-RJ) AND PROCEEDS TO FIND A SOLUTION OF THE DEFLATED POLYNOMIAL, AND SO ON. IF THE COMPUTED SOLUTION RJ IS COMPLEX, THE QUADRATIC FACTOR (X-RJ)(X-CONJG(RJ)) IS USED TO DEFLATE. THE PROCESS CONTINUES UNTIL THE STAGE WHERE THE DEFLATED POLYNOMIAL IS OF DEGREE <= 1 IS REACHED.

FOR INCREASED ACCURACY, EACH COMPUTED SOLUTION RJ IS "PURIFIED" BY USING IT AS AN INITIAL GUESS IN AN ITERATION WITH THE ORIGINAL POLYNOMIAL P(X).

ALL ITERATIONS WITH THE DEFLATED POLYNOMIALS START WITH THE INITIAL GUESS X0 = 0.0. FOR EACH ITERATION, THE ROUTINE COMPUTES SUCCESSIVE ITERATES UNTIL ONE OF THE FOLLOWING TERMINATION CONDITIONS IS SATISIFIED:

  1. |XR - XRM1| <= 4*EPS*(1+|XR|),
      WHERE XR,XRM1 ARE THE TWO MOST RECENT ITERATES
      AND EPS IS MACHINE EPSILON;

  2. |PD(XR)| <= 4*EPS,
      WHERE PD(X) IS THE CURRENT DEFLATED POLYNOMIAL;

  3. 50 ITERATIONS, INCLUDING PUFIFICATION, WERE
      PERFORMED IN AN ATTEMPT TO FIND A ROOT AND
      CONVERGENCE WAS NOT ACHIEVED.

THE ROUTINE INCLUDES A FACILITY TO OUTPUT INTERMEDIATE RESULTS. THIS FACILITY IS CONTROLLED THROUGH THE PARAMETER LIST (SEE BELOW).

CALLING SEQUENCE: CALL LAGR(N,A,XREAL,XIMAG,IND,LIST)

PARAMETERS:

ON ENTRY:

    N   - INTEGER
              THE DEGREE OF P(X). N MUST BE <= 20.

           A   - REAL(N+1)
                      THE COEFFICIENTS OF P(X).

           LIST   - INTEGER
                      A PARAMETER USED TO CONTROL THE OUTPUT OF
                      INTERMEDIATE RESULTS:
                      = 0 - NO OUTPUT;
                      = K - OUTPUT EVERY K-TH ITERATION.

                      THE OUTPUT CONSISTS OF TWO COLUMNS:

                        #   -   THE ITERATION NUMBER. THE BEGINNING
                                OF THE PURIFICATION ITERATION IS
                                INDICATED BY RESTARTING THE COUNT
                                AT 1.

                        XR -   THE CURRENT ITERATE.

   ON RETURN:

           XREAL,   - REAL(N)
            XIMAG   THE REAL AND IMAGINARY PARTS OF THE N
                      SOLUTIONS. (THE KTH SOLUTION IS XREAL(K) +
                      I*XIMAG(K), WHERE I = CSQRT(-1).)

           IND   - INTEGER(N)
                      THE VALUE OF IND(I) INDICATES THE STATUS OF
                      THE I-TH SOLUTION THAT WAS FOUND:
                      = 1   TERMINATION CONDITION (1) IS SATISFIED;
                      = 2   TERMINATION CONDITION (2) IS SATISFIED;
                      = 12 BOTH TERMINATION CONDITIONS (1) AND (2)
                            ARE SATISFIED;
                      = 3   CONVERGENCE COULD NOT BE ACHIEVED WITHIN
                            50 ITERATIONS. NO FURTHER COMPUTATIONS
                            WERE DONE.

           LIST   - THE NUMBER OF SOLUTIONS FOUND.

   SAMPLE CALLING PROGRAM:

         REAL A(21),XREAL(20),XIMAG(20)
         COMPLEX B(21),ROOT
         INTEGER IND(20)
         READ(5,*) N
         NP1 = N + 1
         READ(5,*) (A(I),I=1,NP1)
         LIST = 1
         CALL LAGR(N,A,XREAL,XIMAG,IND,LIST)
   C
   C   *************************
   C   *
   C   *   COMPUTE RESIDUALS AND OUTPUT RESULTS.
   C   *
   C   *************************

```fortran
C
      WRITE(6,1)
      DO 10 K = 1, LIST
         DO 5 I = 1,NP1
            B(I) = CMPLX(A(I),0.0)
 5       CONTINUE
         ROOT = CMPLX(XREAL(K),XIMAG(K))
         CALL CLDF(NP1,B,ROOT)
         WRITE(6,2)  XREAL(K), XIMAG(K), B(NP1), IND(K)
 10   CONTINUE
      STOP
 1    FORMAT (///17X,1HX,30X,4HP(X),14X,6HSTATUS/)
 2    FORMAT (2H (,E14.6,1H,,E14.6,3H)   ,
     +        1H(,E14.6,1H,,E14.6,5H)     ,I2)
      END
```

```
      SUBROUTINE LDF(N,A,R)
```

THIS ROUTINE DIVIDES A GIVEN REAL POLYNOMIAL OF DEGREE N-1,

```
      P(X) = A(1)*X**(N-1) + ... + A(N-1)*X + A(N)
```

BY THE LINEAR TERM (X-R), WHERE R IS REAL. THAT IS, IT CALCULATES THE POLYNOMIAL Q(X), OF DEGREE <= N-2, AND THE CONSTANT REM SUCH THAT

```
      P(X) = (X-R)*Q(X) + REM.
```

THE COEFFICIENTS OF Q ARE STORED IN THE ARRAY A SO THAT

```
      Q(X) = A(1)*X**(N-2) + ... + A(N-2)*X + A(N-1).
```

THE REMAINDER REM IS STORED IN A(N).

THIS ROUTINE CAN BE USED FOR DEFLATION IN CONJUNCTION WITH THE ROOT-FINDING ROUTINES BSECT,SECNT,NEWTN AND ZERO.

THE ROUTINE CAN ALSO BE USED FOR EVALUATING P(X) AT X = R, THAT IS,

```
      P(R) = REM    ( = A(N) ).
```

NOTE THAT, SINCE THE ARRAY A IS OVERWRITTEN, IT IS ADVISABLE TO MAKE A COPY OF THE COEFFICIENTS BEFORE USING THE ROUTINE FOR THIS PURPOSE.

CALLING SEQUENCE: CALL LDF(N,A,R)

PARAMETERS:

ON ENTRY:

      N   - INTEGER
              N-1 IS THE DEGREE OF P(X).

      A   - REAL(N)
              THE COEFFICIENTS OF P(X).

      R   - REAL
              IT FORMS THE DIVISOR (X-R).

ON RETURN:

      A   - THE COEFFICIENTS OF Q(X) WITH THE
              REMAINDER IN A(N).

```
      SUBROUTINE CLDF(N,A,R)
```

THIS ROUTINE DIVIDES A GIVEN COMPLEX POLYNOMIAL, OF DEGREE N-1,

$$P(X) = A(1)*X**(N-1) + \ldots + A(N-1)*X + A(N),$$

BY A LINEAR TERM (X-R). THAT IS, IT CALCULATES THE POLYNOMIAL Q(X), OF DEGREE <= N-2, AND THE CONSTANT REM SUCH THAT

$$P(X) = (X-R)*Q(X) + REM$$

THE COEFFICIENTS OF Q ARE STORED IN THE ARRAY A SO THAT

$$Q(X) = A(1)*X**(N-2) + \ldots + A(N-2)*X + A(N-1).$$

THE REMAINDER REM IS STORED IN A(N).

THIS ROUTINE CAN BE USED FOR DEFLATION IN CONJUNCTION WITH THE ROOT-FINDING SUBROUTINE MULR.

THE ROUTINE CAN ALSO BE USED FOR EVALUATING P(X) AT X = R, THAT IS,

$$P(R) = REM \quad (= A(N)).$$

NOTE THAT, SINCE THE ARRAY A IS OVERWRITTEN IT IS ADVISABLE TO MAKE A COPY OF THE COEFFICIENTS BEFORE USING THE ROUTINE FOR THIS PURPOSE.

CALLING SEQUENCE: CALL CLDF(N,A,R)

PARAMETERS:

ON ENTRY:

       N   - INTEGER
                 N-1 IS THE DEGREE OF P(X).

       A   - COMPLEX(N)
                 THE COEFFICIENTS OF P(X).

       R   - COMPLEX
                 IT FORMS THE DIVISOR (X-R).

ON RETURN:

       A   - THE COEFFICIENTS OF Q(X) WITH THE
                 REMAINDER IN A(N).

```
      SUBROUTINE CQDF(N,A,R)
```

THIS ROUTINE DIVIDES A GIVEN COMPLEX POLYNOMIAL OF DEGREE N-1,

    P(X) = A(1)*X**(N-1) + ... + A(N-1)*X + A(N)

BY THE QUADRATIC TERM (X-R)*(X-RCONJ), WHERE R IS A COMPLEX NUMBER. THAT IS, IT CALCULATES THE POLYNOMIAL Q(X), OF DEGREE <= N-3, AND THE POLYNOMIAL REM(X) OF DEGREE <= 1, SUCH THAT

    P(X) = ((X-R)*(X-RCONJ))*Q(X) + REM(X)

THE COEFFICIENTS OF Q ARE STORED IN THE ARRAY A SO THAT

    Q(X) = A(1)*X**(N-2) + ... + A(N-3)*X + A(N-2).

THE COEFFICIENTS OF REM ARE STORED IN A(N-1) AND A(N) SO THAT

    REM(X) = A(N-1)*X + A(N).

CALLING SEQUENCE: CALL CQDF(N,A,R)

PARAMETERS:

ON ENTRY:

    N   - INTEGER
             N-1 IS THE DEGREE OF P(X).

    A   - COMPLEX(N)
             THE COEFFICIENTS OF P(X).

    R   - COMPLEX
             IT FORMS THE DIVISOR (X-R)*(X-RCONJ).

ON RETURN:

    A   - THE COEFFICIENTS OF Q(X) WITH THE REMAINDER REM(X) IN A(N-1) AND A(N).

THE FOLLOWING PROGRAM IS AN EXAMPLE OF HOW THIS ROUTINE COULD BE USED IN CONJUNCTION WITH THE ROUTINES MULR AND CLDF TO FIND THE ROOTS OF A POLYNOMIAL P(X) WITH REAL COEFFICIENTS. NOTE THAT N AND THE ARRAY A WHICH DEFINE THE POLYNOMIAL ARE PLACED IN COMMON. THIS IS DONE IN ORDER TO ALLOW ACCESS BY CLDF, CQDF AND THE USER-SUPPLIED FUNCTION SUBPROGRAM, F(X), FOR EVALUATING THE POLYNOMIAL.

SAMPLE CALLING PROGRAM USING MULLER'S METHOD:

```
      COMPLEX A(10),X0
      EXTERNAL F
      COMMON A,N
      READ (5,*) N,(A(I),I=1,N)
      READ (5,*) X0
      XTOL = 1.E-5
      FTOL = 1.E-5
      LIST = 0
      RTEST = 1.E-20
      WRITE (6,*) 'NO. ITERATIONS    --ROOTS--'
C
C     *************************
C     *
C     *   MAIN LOOP TO FIND ROOTS.
C     *
C     *   THE INITIAL GUESS X0 USED IN EACH CALL (EXCEPT THE
C     *   FIRST) TO MULR IS THE PREVIOUS ROOT THAT WAS FOUND.
C     *   COMPUTATIONS ARE TERMINATED ON AN ABNORMAL RETURN FROM
C     *   MULR (IND = 3).
C     *
C     *************************
C
10    IF (N .LE. 1) STOP
         ITMAX = 20
         CALL MULR(F,X0,XTOL,FTOL,ITMAX,IND,LIST)
         IF (IND .NE. 3) GO TO 20
            WRITE (6,*) 'MAX ITERATIONS.  LAST APPROX IS ',X0
            STOP
C
C     *************************
C     *
C     *   IF A COMPLEX ROOT IS FOUND, DEFLATE USING CQDF.
C     *   OTHERWISE USE CLDF.
C     *
C     *************************
C
20       IF (ABS(AIMAG(X0)) .LE. RTEST) GO TO 30
            WRITE (6,*) ITMAX,'   ',X0,CONJG(X0)
            CALL CQDF(N,A,X0)
            N = N - 2
            GO TO 10
30       WRITE (6,*) ITMAX,'   ',X0
         CALL CLDF(N,A,X0)
         N = N - 1
         GO TO 10
      END

      COMPLEX FUNCTION F(X)
      COMPLEX X,A(10)
      COMMON  A,N
      F = A(1)
      DO 10 K = 2,N
```

```
          F = A(K) + X*F
10     CONTINUE
       RETURN
       END
```

```
SUBROUTINE BRYDN(FCN,N,XM1,X0,XTOL,FTOL,ITMAX,IND,LIST)
```

THIS SUBROUTINE USES BROYDEN'S METHOD TO FIND A SOLUTION OF
THE SYSTEM OF N NONLINEAR EQUATIONS IN N UNKNOWNS

$$F_I(X) = F_I(X(1),...X(N)) = 0, \quad 1 <= I <= N.$$

GIVEN TWO INITIAL GUESSES XM1 AND X0, THE ROUTINE COMPUTES
SUCCESSIVE ITERATES XR = (XR(1),...,XR(N)) UNTIL ONE OF THE
FOLLOWING TERMINATION CONDITIONS IS SATISFIED:

1. $|XR(J)-XRM1(J)|$ <= XTOL*(1.0+$|XR(J)|$), 1 <= J <= N,
   WHERE XRM1 AND XR ARE THE TWO MOST RECENT ITERATES;

2. $|F_I(XR)|$ <= FTOL, 1 <= I <= N;

3. ITMAX ITERATIONS HAVE BEEN PERFORMED.

CALLING SEQUENCE: BRYDN(FCN,N,XM1,X0,XTOL,FTOL,ITMAX,IND,LIST)

PARAMETERS:

ON ENTRY:

    FCN   - SUBROUTINE NAME
             A USER-SUPPLIED SUBROUTINE, FCN(N,X,F), FOR
             EVALUATING EACH OF THE FUNCTIONS FI(X) AT ANY
             POINT X = (X(1),...,X(N)).

    N   - INTEGER
             THE NUMBER OF EQUATIONS IN THE SYSTEM.

    XM1,   - REAL(N)
     X0   INITIAL GUESSES FOR THE SOLUTION.

    XTOL   - REAL
             THE ACCEPTABLE TOLERANCE IN THE COMPONENT-
             WISE CHANGES $|XR(J) - XRM1(J)|$, J = 1,...,N,
             BETWEEN SUCCESSIVE ITERATES.

    FTOL   - REAL
             THE ACCEPTABLE TOLERANCE IN THE FUNCTION
             VALUES $|F_I(XR)|$, I = 1,...,N.

    ITMAX   - INTEGER
             THE MAXIMUM NUMBER OF ITERATIONS TO BE
             PERFORMED.

    LIST   - INTEGER

```
                        A  PARAMETER  USED  TO  CONTROL   OUTPUT  OF
                        INTERMEDIATE RESULTS:
                        = 0   - NO OUTPUT;
                          K   - OUTPUT EVERY K-TH ITERATION.

    ON RETURN:

        XM1,    -  THE  TWO  MOST  RECENT ITERATES. THE APPROX-
         X0        IMATE  SOLUTION  IS  X0  SINCE IT IS  THE  MOST
                   RECENT ITERATE.

        ITMAX   -  THE NUMBER OF ITERATIONS PERFORMED.

        IND     -  INTEGER
                   THE STATUS OF THE RESULT:
                   = 1    TERMINATION CONDITION (1) IS SATISFIED;
                   = 2    TERMINATION CONDITION (2) IS SATISFIED;
                   = 12   BOTH TERMINATION CONDITIONS (1) AND (2)
                            ARE SATISFIED;
                   = 3    CONVERGENCE   COULD   NOT   BE  ACHIEVED
                            WITHIN ITMAX ITERATIONS.

    SAMPLE CALLING PROGRAM:

          EXTERNAL FCN
          DIMENSION XM1(10),X0(10)
          READ(5,*) N, (XM1(J),J=1,N), (X0(J),J=1,N)
          XTOL = 1.E-5
          FTOL = 1.E-5
          ITMAX = N + 30
          LIST = 0
          CALL BRYDN(FCN,N,XM1,X0,XTOL,FTOL,ITMAX,IND,LIST)
          WRITE (6,*) '# ITERATIONS = ',ITMAX,'IND = ',IND
          WRITE (6,*) 'SOLUTION:'
          DO 19 I = 1,N
              WRITE (6,*) I,XM1(I),I,X0(I)
    19    CONTINUE
          STOP
          END

          SUBROUTINE FCN(N,X,F)
          DIMENSION  X(N),F(N)
            ... EXPRESSIONS FOR EVALUATING EACH FI(X)
          RETURN
          END


    REFERENCE:              /
             DENNIS AND MORE
             SIAM REVIEW, JAN 1977
```

```
      SUBROUTINE SNWTN(FCN,JACOB,N,XO,XTOL,FTOL,ITMAX,IND,LIST,AJ)
```

THIS SUBROUTINE USES NEWTON'S METHOD TO FIND A SOLUTION OF THE SYSTEM OF N NONLINEAR EQUATIONS IN N UNKNOWNS

$$F_I(X(1),\ldots X(N)) = 0, \quad 1 <= I <= N.$$

GIVEN AN INITIAL GUESS XO = (XO(1),...XO(N)), THE ROUTINE COMPUTES SUCCESSIVE ITERATES XR = (XR(1),...XR(N)) UNTIL ONE OF THE FOLLOWING TERMINATION CONDITIONS IS SATISFIED:

1. |XR(J)-XRM1(J)| <= XTOL*(1.0+|XR(J)|), 1 <= J <= N, WHERE XRM1 AND XR ARE THE TWO MOST RECENT ITERATES;

2. $|F_I(XR)|$ <= FTOL, 1 <= I <= N;

3. ITMAX ITERATIONS HAVE BEEN PERFORMED;

4. THE JACOBIAN MATRIX AJ(XR) IS NUMERICALLY SINGULAR (AS DEFINED IN THE DOCUMENTATION FOR THE SUB- ROUTINE DCOMP).

CALLING SEQUENCE: CALL SNWTN(FCN,JACOB,N,XO,XTOL,FTOL,ITMAX, IND,LIST,AJ)

PARAMETERS:

ON ENTRY:

    FCN   - SUBROUTINE NAME
              A USER-SUPPLIED SUBROUTINE, FCN(N,X,F), FOR EVALUATING EACH OF THE FUNCTIONS FI(X) AT ANY POINT X = (X(1),...,X(N)).

    JACOB   - SUBROUTINE NAME
              A USER-SUPPLIED SUBROUTINE, JACOB(N,X,AJ), FOR EVALUATING THE JACOBIAN MATRIX OF THE SYSTEM (RETURNED IN AJ).

    N   - INTEGER
              THE NUMBER OF EQUATIONS IN THE SYSTEM. N MUST BE <= 20.

    XO   - REAL(N)
              THE INITIAL GUESS FOR THE SOLUTION.

    XTOL   - REAL
              THE ACCEPTABLE TOLERANCE IN THE COMPONENT- WISE CHANGES |XR(J) - XRM1(J)|, J = 1,...,N, BETWEEN SUCCESSIVE ITERATES.

FTOL   - REAL
           THE ACCEPTABLE TOLERANCE IN THE FUNCTION
           VALUES $|F_I(XR)|$, $I = 1,\ldots,N$.

ITMAX   - INTEGER
           THE MAXIMUM NUMBER OF ITERATIONS TO BE
           PERFORMED.

LIST   - INTEGER
           A PARAMETER USED TO CONTROL OUTPUT OF
           INTERMEDIATE RESULTS:
           = 0 - NO OUTPUT;
           = K - OUTPUT EVERY K-TH ITERATION.

AJ   - REAL(N,N)
           WORKSPACE. IT IS USED FOR HANDLING THE
           JACOBIAN MATRIX OF THE SYSTEM. THIS
           PARAMETER NEED NOT BE INITIALIZED.

ON RETURN:

XO   - THE MOST RECENT APPROXIMATE SOLUTION.

ITMAX   - THE NUMBER OF ITERATIONS PERFORMED.

IND   - INTEGER
           THE STATUS OF THE RESULT:
           = 1   TERMINATION CONDITION (1) IS SATISFIED;
           = 2   TERMINATION CONDITION (2) IS SATISFIED;
           = 12 BOTH TERMINATION CONDITIONS (1) AND (2)
                 ARE SATISFIED;
           = 3   CONVERGENCE COULD NOT BE ACHIEVED
                 WITHIN ITMAX ITERATIONS;
           = 4   THE JACOBIAN IS NUMERICALLY SINGULAR.

SAMPLE CALLING PROGRAM:

```
      DIMENSION X(20),F(20),AJ(20,20)
      EXTERNAL FCN,JACOB
        READ (5,*) N,(X(I),I=1,N)
        XTOL = 1.E-5
        FTOL = 1.E-5
        ITMAX = N + 15
        LIST = 0
        CALL SNWTN(FCN,JACOB,N,X,XTOL,FTOL,ITMAX,IND,LIST,AJ)
        WRITE (6,*) '# ITERATIONS = ',ITMAX,'IND = ',IND
        WRITE (6,*) 'SOLUTION:'
        DO 19 I = 1,N
             WRITE (6,*) I,X(I)
   19   CONTINUE
      STOP
      END
```

```
      SUBROUTINE FCN(N,X,F)
      DIMENSION  X(N),F(N)
         ... EXPRESSIONS FOR EVALUATING EACH FI(X)
      RETURN
      END


      SUBROUTINE JACOB(N,X,AJ)
      DIMENSION AJ(N,N),X(N)
         ... EXPRESSIONS FOR EVALUATING EACH COMPONENT
             OF THE JACOBIAN MATRIX:   AJ(I,J) = D(FI)/D(XJ)
      RETURN
      END
```

```
      FUNCTION QUAD(RULE,N,F,A,B)
```

THIS FUNCTION SUBPROGRAM APPROXIMATES THE VALUE OF AN INTEGRAL I(F;A,B) OF A FUNCTION F(X) OVER THE INTERVAL (A,B) USING THE N-POINT QUADRATURE RULE

$$Q(I;A,B) = W_1 * F(X_1) + W_2 * F(X_2) + \ldots + W_N * F(X_N)$$

THE PARTICULAR RULE (DETERMINING THE QUADRATURE POINTS XI AND THE WEIGHTS WI) TO BE USED IS SPECIFIED BY THE PARAMETERS RULE AND N.

CALLING SEQUENCE: APPROX = QUAD(RULE,N,F,A,B)

PARAMETERS:

ON ENTRY:

    RULE   - SUBROUTINE NAME
              A ROUTINE, RULE(N,A,B,P,W), FOR SUPPLYING THE QUADRATURE POINTS XI AND THE WEIGHTS WI RELATIVE TO THE INTERVAL OF INTEGRATION (A,B). THE FOLLOWING VALUES WILL ACCESS THE RULES THAT HAVE BEEN IMPLEMENTED IN THE PACKAGE:
                RULE = NCQ   - NEWTON-COTES RULES;
                     = GLEGQ - GAUSS-LEGENDRE RULES;
                     = GLAGQ - GAUSS-LAGUERRE RULES.
              SEE THE DOCUMENTATION OF THE APPROPRIATE SUBROUTINE FOR DETAILS.
              IN ADDITION, A USER-SUPPLIED SUBROUTINE IMPLEMENTATION OF A QUADRATURE RULE CAN BE USED. SUCH A ROUTINE MUST, OF COURSE, CONFORM TO THE ABOVE CALLING SEQUENCE.

    N   - INTEGER
              THE NUMBER OF QUADRATURE POINTS TO BE USED. THE LIMITS ON N WHEN RULE =
                NCQ   : 1 <= N <= 21;
                GLEGQ : 2 <= N <= 20;
                GLAGQ : 2 <= N <= 10.

    F   - REAL
              A USER-SUPPLIED FUNCTION SUBPROGRAM, F(X), FOR EVALUATING THE INTEGRAND AT ANY POINT X IN (A,B).

    A,B   - REAL
              THE LEFT AND RIGHT ENDPOINTS OF THE INTERVAL OF INTEGRATION.

ON RETURN:

```
          QUAD    - REAL
                   THE APPROXIMATE VALUE OF I(F;A,B).


   SAMPLE CALLING PROGRAM:

C     *************************
C     *
C     *   THIS PROGRAM APPROXIMATES THE VALUE OF I(F;A,B) USING
C     *   A COMPOSITE 5-POINT GAUSS-LEGENDRE RULE.  THE NUMBER
C     *   OF SUBDIVISIONS OF (A,B) IS MDIV.
C     *
C     *************************
C
      REAL A,B,H,XR,XL,APPROX
      EXTERNAL F,GLEGQ
      READ (5,*) A,B,MDIV
      H = (B - A)/FLOAT(MDIV)
      XR = A
      APPROX = 0.E0
      DO 10 IDIV = 1, MDIV
           XL = XR
           XR = A + FLOAT(IDIV)*H
           APPROX = APPROX + QUAD(GLEGQ,5,F,XL,XR)
10    CONTINUE
      WRITE (6,*) 'APPROX. VALUE OF THE INTEGRAL = ',APPROX
      STOP
      END

      FUNCTION F(X)
      F = ... EXPRESSION TO EVALUATE THE INTEGRAND
      RETURN
      END
```

```
FUNCTION AQUAD(RULE,N,F,A,B,TOL,MAXLEV,SING,ERREST)
```

THIS FUNCTION SUBPROGRAM APPROXIMATES THE VALUE OF A DEFINITE INTEGRAL I(F;A,B) OF A FUNCTION F(X) OVER THE INTERVAL (A,B) USING AN ADAPTIVE QUADRATURE PROCEDURE. THE UNDERLYING QUADRATURE RULE TO BE USED IS SPECIFIED BY THE PARAMETERS RULE AND N.

THE ROUTINE SUCCESSIVELY BISECTS THE INTERVAL OF INTEGRATION AND APPLIES THE SPECIFIED QUADRATURE RULE OVER EACH SUBINTERVAL. A SUBINTERVAL (XL,XR) IS "ACCEPTED" IF

```
   |Q(F;XL,XR) - ( Q(F;XL,XM) + Q(F;XM,XL) )|

                           <= TOL*FAC*|XR - XL|/|B - A|,
```

WHERE
  Q(F;XL,XR) DENOTES THE APPROXIMATE VALUE OF THE INTEGRAL
    I(F;XL,XR), OF F(X) OVER (XR,XL);
  XM IS THE MIDPOINT OF (XL,XR), AND
  FAC IS A SCALING FACTOR, DEPENDENT ON N AND THE TYPE OF
    RULE. IN PARTICULAR,
      FAC = 2**(N+2)-1   FOR NEWTON-COTES RULES,
          = 2**(2*N)-1   FOR GAUSS-LEGENDRE RULES.
    THE CORRECT VALUE OF FAC IS PROVIDED AUTOMATICALLY BY
    THE SUBROUTINE SPECIFIED BY THE VALUE OF RULE.

IF THIS TEST FAILS, THEN (XL,XR) IS BISECTED AND THE PROCEDURE IS REPEATED ON THE LEFT HALF. WHEN A SUBINTERVAL IS ACCEPTED, THE ROUTINE GOES TO THE NEXT SUBINTERVAL TO THE RIGHT, AND SO ON.

THE MAXIMUM LEVEL OF BISECTION CAN BE CONTROLLED THROUGH THE PARAMETER MAXLEV. IF THE MAXIMUM LEVEL OF BISECTION HAS BEEN REACHED FOR A SUBINTERVAL (XL,XR) BUT THE ACCEPTANCE TEST STILL FAILS THEN THE MIDPOINT XM IS RECORDED IN "SING" AND THE ROUTINE WILL CONTINUE THE INTEGRATION WITH THE NEXT SUBINTERVAL TO THE RIGHT. THE VALUE OF THE INTEGRAL OVER THE "FAILED" SUBINTERVAL IS TAKEN TO BE Q(F;XL,XR).

CALLING SEQUENCE: APPROX = AQUAD(RULE,N,F,A,B,TOL,MAXLEV,
                                   SING,ERREST)

PARAMETERS:

ON ENTRY:

        RULE   - SUBROUTINE NAME
                  A ROUTINE, RULE(N,XL,XR,P,W), FOR SUPPLYING
                  THE QUADRATURE POINTS XI AND THE WEIGHTS WI
                  RELATIVE TO THE INTERVAL (XL,XR) OF
                  INTEGRATION. THE FOLLOWING VALUES WILL

ACCESS THE RULES THAT HAVE BEEN IMPLEMENTED
IN THE PACKAGE:
   RULE = NCQ   - NEWTON-COTES RULES;
        = GLEGQ   - GAUSS-LEGENDRE RULES.
NOTE: SINCE GAUSS-LAGUERRE RULES ARE NOT
SUITABLE FOR ADAPTIVE QUADRATURE, THE ROUTINE
GLAGQ SHOULD NOT BE USED WITH AQUAD.
A USER-SUPPLIED SUBROUTINE IMPLEMENTATION OF
A QUADRATURE RULE CAN ALSO BE USED. SUCH A
ROUTINE MUST, OF COURSE, CONFORM TO THE ABOVE
CALLING SEQUENCE.

N   - INTEGER
          THE NUMBER OF QUADRATURE POINTS TO BE USED
          IN EACH SUBINTERVAL (XL,XR). LIMITS ON N
          WHEN RULE =
             NCQ   : 1 <= N <= 21;
             GLEGQ : 2 <= N <= 20.

F   - REAL
          A USER-SUPPLIED FUNCTION SUBPROGRAM, F(X),
          FOR EVALUATING THE INTEGRAND AT ANY POINT X
          IN (A,B).

A,B   - REAL
          THE LEFT AND RIGHT ENDPOINTS OF THE INTERVAL
          OF INTEGRATION.

TOL   - REAL
          USED IN THE CRITERION FOR ACCEPTING A
          SUBINTERVAL.

MAXLEV - INTEGER
          THE MAXIMUM LEVEL OF BISECTION ALLOWED. THE
          SMALLEST POSSIBLE LENGTH OF A SUBINTERVAL
          WILL BE (B - A)/2**MAXLEV, WHERE
          0 <= MAXLEV <= 30.

ON RETURN:

AQUAD   - REAL
          THE APPROXIMATE VALUE OF THE INTEGRAL.

TOL   - THE FRACTION OF THE INTERVAL OF INTEGRATION
          (A,B) ON WHICH THE SUBINTERVALS WERE NOT
          ACCEPTED. IF TOL > 0, THERE HAVE BEEN SOME
          DIFFICULTIES. THE USER SHOULD CHECK WHETHER
          - THE INTEGRAND HAS A SINGULARITY
            (AT X = SING), OR ELSE
          - THE INPUT VALUES OF TOL AND/OR MAXLEV WERE
            TOO SMALL FOR ACCEPTANCE TO OCCUR BEFORE
            THE MAXIMUM NUMBER OF BISECTIONS WAS DONE.

MAXLEV - THE NUMBER OF SUBINTERVALS INTO WHICH THE
          INTERVAL OF INTEGRATION (A,B) WAS DIVIDED.

```
            SING    - REAL
                      WHEN TOL > 0, SING WILL BE THE MIDPOINT OF
                      THE SUB-INTERVAL THAT IS CLOSEST TO B AND
                      FOR WHICH THE ACCEPTANCE CRITERION FAILED
                      AFTER MAXLEV LEVELS OF BISECTION WERE DONE.

            ERREST  - REAL
                      AN ESTIMATE OF THE ERROR IN THE VALUE OF
                      AQUAD AS AN APPROXIMATION FOR I(F;A,B).


      SAMPLE CALLING PROGRAM:

C     *************************
C     *
C     *   THIS PROGRAM APPROXIMATES THE VALUE OF I(F;A,B) USING
C     *   ADAPTIVE QUADRATURE BASED ON A 6-POINT GAUSS-LEGENDRE
C     *   RULE.
C     *
C     *************************
C
      REAL A,B,TOL,APPROX,SING,ERREST
      EXTERNAL F,GLEGQ
      READ (5,*) A,B
      TOL = 1.E-5
      MAXLEV = 30
      APPROX = AQUAD(GLEGQ,6,F,A,B,TOL,MAXLEV,SING,ERREST)
      IF (TOL .NE. 0.E0) WRITE (6,*) 'ROUTINE FAILED. TOL = ',TOL,
     +                               '  SING = ',SING
      WRITE (6,*) 'APPROXIMATE VALUE OF INTEGRAL = ',APPROX,
     +            ' ERROR ESTIMATE = ',ERREST,
     +            ' # SUBINTERVALS USED = ',MAXLEV
      STOP
      END

      FUNCTION F(X)
      F = ... EXPRESSION TO EVALUATE THE INTEGRAND
      RETURN
      END
```

```
      FUNCTION ROM(F,A,B,MDIV,TOL,MAXROW,IND,LIST)
```

THIS FUNCTION SUBPROGRAM APPROXIMATES THE VALUE OF THE DEFINITE INTEGRAL I(F;A,B) OF A FUNCTION F(X) OVER A SPECIFIED INTERVAL (A,B) USING ROMBERG EXTRAPOLATION WITH INTERVAL HALVING.

THE ENTRIES IN THE ROMBERG TABLE ARE

```
        T(K,M),   0 <= M <= K <= MAXROW-1.
```

THE ROUTINE WILL STOP WHEN ROW CONVERGENCE ON TWO SUCCESSIVE ROWS IS ACHIEVED, THAT IS, WHEN

```
        !RDEL(K-1,M)! <= TOL*(1 + !T(K-1,M)!), AND
        !RDEL(K,M)!   <= TOL*(1 + !T(K,M)!),
```

WHERE

```
        RDEL(I,J) = T(I,J) - T(I,J-1).
```

CALLING SEQUENCE: APPROX = ROM(F,A,B,MDIV,TOL,MAXROW,IND,LIST)

PARAMETERS:

ON ENTRY:

  F   - REAL
      A USER-SUPPLIED FUNCTION SUBPROGRAM, F(X), FOR EVALUATING THE INTEGRAND AT ANY POINT X IN (A,B).

  A,B   - REAL
      THE LEFT AND RIGHT ENDPOINTS OF THE INTERVAL OF INTEGRATION.

  MDIV - INTEGER
      THE NUMBER OF SUB-DIVISIONS INTO WHICH THE INTERVAL (A,B) IS TO BE DIVIDED AT THE OUTSET.

  TOL   - REAL
      THE TOLERANCE FOR THE CONVERGENCE CRITERION.

  MAXROW - INTEGER
      THE MAXIMUM NUMBER OF ROWS IN THE ROMBERG TABLE THAT ARE TO BE COMPUTED.
      RESTRICTION: 3 <= MAXROW <= 20.

  LIST - INTEGER
      A PARAMETER USED TO CONTROL OUTPUT OF INTERMEDIATE RESULTS:

```
                        = 0  - NO OUTPUT;
                        = 1  - THE ROMBERG TABLE WILL BE OUTPUT.

    ON RETURN:

        ROM     - REAL
                    THE APPROXIMATE VALUE OF I(F;A,B).

        IND     - INTEGER
                    THE STATUS OF THE APPROXIMATE VALUE:
                    = 0  - CONVERGENCE WAS ACHIEVED;
                    = 1  - CONVERGENCE WAS NOT ACHIEVED WITHIN
                           ITMAX ROWS.  THE LAST ENTRY IN THE
                           IN THE TABLE WAS RETURNED.


    SAMPLE CALLING PROGRAM:

          REAL A,B,TOL,APPROX
          EXTERNAL F
          READ (5,*) A,B,MDIV
          TOL = 1.E-5
          MAXROW = 20
          LIST = 0
          APPROX = ROM(F,A,B,MDIV,TOL,MAXROW,IND,LIST)
          IF (IND .EQ. 0) GO TO 10
               WRITE (6,*) 'NO CONVERGENCE.   LAST ENTRY IN TABLE IS ',
         *                  APPROX
               STOP
    10    WRITE (6,*) 'I(F;A,B) = ',APPROX
          STOP
          END

          FUNCTION F(X)
          F = ... EXPRESSION TO EVALUATE THE INTEGRAND
          RETURN
          END
```

FUNCTION CROM(F,A,B,MDIV,TOL,MAXROW,MAXCOL,IND,LIST)

THIS FUNCTION SUBPROGRAM APPROXIMATES THE VALUE OF THE DEFINITE INTEGRAL I(F;A,B) OF A FUNCTION F(X) OVER A GIVEN INTERVAL (A,B) USING CAUTIOUS ROMBERG EXTRAPOLATION WITH INTERVAL HALVING.

THE ENTRIES IN THE ROMBERG TABLE ARE

T(K,M),   (0 <= K <= MAXROW-1,
          (
          (0 <= M <= MIN(K,MAXCOL-1).

ROW AND COLUMN DIFFERENCES ARE DENOTED, RESPECTIVELY, BY

RDEL(K,M) = T(K,M) - T(K,M-1),
CDEL(K,M) = T(K,M) - T(K-1,M).

THE ROUTINE WILL STOP WHEN ROW CONVERGENCE ON TWO SUCCESSIVE ROWS IS ACHIEVED, THAT IS, WHEN

!RDEL(K-1,M)! <= TOL*(1 + !T(K-1,M)!), AND
!RDEL(K,M)! <= TOL*(1 + !T(K,M)!).

IN CASES WHEN THE MAXIMUM NUMBER OF EXTRAPOLATIONS HAS BEEN TAKEN OR WHEN EXTRAPOLATION IS NOT VALID, COLUMN CONVERGENCE IS CHECKED, I.E.,

!CDEL(K-1,M)! <= TOL*(1 + !T(K-1,M)!), AND
!CDEL(K,M)! <= TOL*(1 + !T(K,M)!).

EXTRAPOLATION TO T(K,M+1) WILL BE DONE (IS FOUND TO BE VALID) IF

$$\left| \, |CDEL(K-1,M)/CDEL(K,M)| - 4^M \, \right| \leq 4^M * 1.E-1.$$

CALLING SEQUENCE: APPROX = CROM(F,A,B,MDIV,TOL,MAXROW,MAXCOL,
                                  IND,LIST)

PARAMETERS:

ON ENTRY:

    F   - REAL
               A USER-SUPPLIED FUNCTION SUBPROGRAM, F(X),
               FOR EVALUATING THE INTEGRAND AT ANY POINT X
               IN (A,B).

    A,B   - REAL
               THE LEFT AND RIGHT ENDPOINTS OF THE INTERVAL
               OF INTEGRATION.

MDIV   - INTEGER
          THE NUMBER OF SUB-DIVISIONS INTO WHICH THE
          INTERVAL (A,B) IS TO BE DIVIDED AT THE
          OUTSET.

TOL   - REAL
          THE TOLERANCE FOR THE CONVERGENCE CRITERIA.

MAXROW - INTEGER
          THE MAXIMUM NUMBER OF ROWS IN THE ROMBERG
          TABLE THAT ARE TO BE COMPUTED.
          RESTRICTION: 3 <= MAXROW.

MAXCOL - INTEGER
          THE MAXIMUM NUMBER OF COLUMNS IN THE ROMBERG
          TABLE THAT ARE TO BE COMPUTED. (THE MAXIMUM
          NUMBER OF EXTRAPOLATIONS WILL THEREFORE BE
          MAXCOL-1.).   RESTRICTION: 2 <= MAXCOL <= 20.

LIST   - INTEGER
          A PARAMETER USED TO CONTROL OUTPUT OF
          INTERMEDIATE RESULTS:
          = 0 - NO OUTPUT;
          = 1 - THE ROMBERG TABLE WILL BE OUTPUT.

ON RETURN:

CROM   - REAL
          THE APPROXIMATE VALUE OF I(F;A,B).

IND   - INTEGER
          THE STATUS OF THE APPROXIMATE VALUE:
          = 0 - ROW CONVERGENCE WAS ACHIEVED;
          = 1 - FURTHER EXTRAPOLATION ALONG THE
                CURRENT ROW WAS NOT VALID BUT COLUMN
                CONVERGENCE WAS ACHIEVED;
          = 2 - THE MAXIMUM NUMBER OF EXTRAPOLATIONS
                WAS TAKEN. COLUMN CONVERGENCE WAS
                ACHIEVED DOWN THE LAST COLUMN;
          = 3 - NO CONVERGENCE. THE TABLE ENTRY
                (MAXROW-1,MAXCOL-1) WAS RETURNED.

SAMPLE CALLING PROGRAM:

```
      REAL A,B,TOL,APPROX
      EXTERNAL F
      READ (5,*) A,B,MDIV
      TOL = 1.E-5
      MAXROW = 30
      MAXCOL = 20
      LIST = 0
      APPROX = CROM(F,A,B,MDIV,TOL,MAXROW,MAXCOL,IND,LIST)
      IF (IND .NE. 4) GO TO 10
```

```
      WRITE (6,*) 'NO CONVERGENCE.   LAST ENTRY IN TABLE IS ',
     *            APPROX
      STOP
10    WRITE (6,*) 'I(F;A,B) = ',APPROX
      STOP
      END

      FUNCTION F(X)
      F = ... EXPRESSION TO EVALUATE THE INTEGRAND
      RETURN
      END
```

```
      FUNCTION QUAD2(RULE,N,F,Y1,Y2,X1,X2)
```

THIS FUNCTION SUBPROGRAM APPROXIMATES THE VALUE OF A DOUBLE INTEGRAL I(I(F;Y1,Y2);X1,X2) OF THE FUNCTION F(X,Y) OVER THE REGION BOUNDED BY Y1(X) <= Y <= Y2(X), X1 <= X <= X2, USING A PRODUCT RULE SPECIFIED BY THE PARAMETERS RULE AND N.

NOTE: IF Y1(X) AND Y2(X) INTERSECT BETWEEN X1 AND X2, THEN QUAD2 WILL RETURN AN ERRONEOUS RESULT.

CALLING SEQUENCE: APPROX = QUAD2(RULE,N,F,Y1,Y2,X1,X2)

PARAMETERS:

ON ENTRY:

    RULE   - SUBROUTINE NAME
              A SUBROUTINE, RULE(N,A,B,P,W), FOR SUPPLYING THE APPROPRIATE POINTS AND WEIGHTS FOR THE (ONE-DIMENSIONAL) QUADRATURE RULE THAT FORMS THE BASIS OF THE PRODUCT RULE. THE FOLLOWING VALUES WILL ACCESS THE RULES THAT HAVE BEEN IMPLEMENTED IN THE PACKAGE.
```
              RULE = NCQ      - NEWTON-COTES RULES;
                   = GLEGQ    - GAUSS-LEGENDRE RULES;
                   = GLAGQ    - GAUSS-LAGUERRE RULES.
```
              IN ADDITION, A USER-SUPPLIED SUBROUTINE IMPLEMENTATION OF A QUADRATURE RULE CAN BE USED. SUCH A ROUTINE MUST, OF COURSE, CONFORM TO THE ABOVE CALLING SEQUENCE.

    N   - INTEGER
              THE NUMBER OF POINTS USED BY THE BASIC QUADRATURE FORMULA. LIMITS ON N WHEN RULE =
```
              NCQ   :  1 <= N <= 21;
              GLEGQ :  2 <= N <= 20;
              GLAGQ :  2 <= N <= 10.
```

    F   - REAL
              A USER-SUPPLIED FUNCTION SUBPROGRAM, F(X), FOR EVALUATING THE INTEGRAND AT ANY POINT X IN (A,B).

    Y1,Y2   - REAL
              USER-SUPPLIED FUNCTION SUBPROGRAMS, Y1(X) AND Y2(X), FOR EVALUATING THE LOWER AND UPPER LIMITS, RESPECTIVELY, OF INTEGRATION WITH RESPECT TO THE Y-VARIABLE.

    X1,X2   - REAL
              THE LOWER AND UPPER LIMITS OF INTEGRATION WITH RESPECT TO THE X-VARIABLE.

ON RETURN:

    QUAD2   - REAL
        THE APPROXIMATE VALUE OF THE DOUBLE INTEGRAL.

SAMPLE CALLING PROGRAM:

```
C     *************************
C     *
C     *  THIS PROGRAM APPROXIMATES A DOUBLE INTEGRAL USING A
C     *  PRODUCT RULE BASED ON A 5-POINT GAUSS-LEGENDRE RULE.
C     *
C     *************************
C
      REAL X1,X2,APPROX
      EXTERNAL F,Y1,Y2,GLEGQ
      READ (5,*) X1,X2
      APPROX = QUAD2(GLEGQ,5,F,Y1,Y2,X1,X2)
      WRITE (6,*) 'THE APPROXIMATE VALUE OF THE INTEGRAL IS ',APPROX
      STOP
      END

      FUNCTION F(X,Y)
      F = ... EXPRESSION TO EVALUATE THE INTEGRAND
      RETURN
      END

      FUNCTION Y1(X)
      Y1 = ... EXPRESSION TO EVALUATE LOWER LIMIT Y1(X)
      RETURN
      END

      FUNCTION Y2(X)
      Y2 = ... EXPRESSION TO EVALUATE UPPER LIMIT Y2(X)
      RETURN
      END
```

        SUBROUTINE NCQ(N,A,B,P,W)

THIS SUBROUTINE IS DESIGNED TO BE USED BY EITHER OF THE ROUTINES QUAD OR AQUAD FOR APPROXIMATING THE VALUE OF THE DEFINITE INTEGRAL I(F;A,B) OF F(X) OVER THE INTERVAL (A,B) USING AN N-POINT NEWTON-COTES FORMULA.

THE PURPOSE OF NCQ IS TO SUPPLY, TO QUAD OR AQUAD, THE N QUADRATURE POINTS AND WEIGHTS FOR INTEGRATING OVER THE INTERVAL (A,B). THE ROUTINE DOES THIS BY TAKING THE N POINTS AND WEIGHTS FOR THE "NORMALIZED" INTERVAL (0,1) AND TRANSFORMING THEM TO (A,B).

THIS ROUTINE CAN ALSO BE USED BY THE ROUTINE QUAD2 TO GENERATE PRODUCT NEWTON-COTES RULES FOR APPROXIMATING INTEGRALS IN 2 DIMENSIONS.

CALLING SEQUENCE: CALL NCQ(N,A,B,P,W)

PARAMETERS:

ON ENTRY:

        N   - INTEGER
                   THE NUMBER OF QUADRATURE POINTS TO BE USED.
                   RESTRICTION: 1 <= N <= 21.
                   EXCEPTION: IF N < 0, THE ROUTINE WILL SIMPLY
                     COMPUTE THE VALUE FAC = 2**(N+2) - 1 AND
                     RETURN IT IN P(1). THIS FEATURE IS USED BY
                     THE SUBROUTINE AQUAD.

        A,B   - REAL
                   THE LEFT AND RIGHT ENDPOINTS OF THE INTERVAL
                   OF INTEGRATION.

ON RETURN:

        P   - REAL(N)
                   THE POINTS FOR THE N-POINT NEWTON-COTES
                   QUADRATURE RULE ON THE INTERVAL (A,B).

        W   - REAL(N)
                   THE WEIGHTS FOR THE N-POINT NEWTON-COTES
                   QUADRATURE RULE.

SAMPLE CALLING PROGRAM:

     SEE ANY ONE OF THE ROUTINES QUAD, AQUAD OR QUAD2.

```
      SUBROUTINE GLEGQ(N,A,B,P,W)
```

THIS SUBROUTINE IS DESIGNED TO BE USED BY EITHER OF THE ROUTINES QUAD OR AQUAD FOR APPROXIMATING THE VALUE OF THE DEFINITE INTEGRAL I(F;A,B) OF F(X) OVER THE INTERVAL (A,B) USING AN N-POINT GAUSS-LEGENDRE FORMULA.

THE PURPOSE OF GLEGQ IS TO SUPPLY, TO QUAD OR AQUAD, THE N QUADRATURE POINTS AND WEIGHTS FOR INTEGRATING OVER THE INTERVAL (A,B). THE ROUTINE DOES THIS BY TAKING THE N POINTS AND WEIGHTS FOR THE "NORMALIZED" INTERVAL (-1,1) AND TRANSFORMING THEM TO (A,B).

THIS ROUTINE CAN ALSO BE USED BY THE ROUTINE QUAD2 TO GENERATE PRODUCT GAUSS-LEGENDRE RULES FOR APPROXIMATING INTEGRALS IN 2 DIMENSIONS.

CALLING SEQUENCE: CALL GLEGQ(N,A,B,P,W)

PARAMETERS:

ON ENTRY:

    N   - INTEGER
            THE NUMBER OF QUADRATURE POINTS TO BE USED.
            RESTRICTION: 2 <= N <= 20.
            EXCEPTION: IF N < 0, THE ROUTINE WILL SIMPLY
              COMPUTE THE VALUE FAC = 2**(2*N) - 1 AND
              RETURN IT IN P(1). THIS FEATURE IS USED BY
              THE SUBROUTINE AQUAD.

    A,B   - REAL
            THE LEFT AND RIGHT ENDPOINTS OF THE INTERVAL
            OF INTEGRATION.

ON RETURN:

    P   - REAL(N)
            THE POINTS FOR THE N-POINT GAUSS-LEGENDRE
            QUADRATURE RULE ON THE INTERVAL (A,B).

    W   - REAL(N)
            THE WEIGHTS FOR THE N-POINT GAUSS-LEGENDRE
            QUADRATURE RULE.

SAMPLE CALLING PROGRAM:

    SEE ANY ONE OF THE ROUTINES QUAD, AQUAD OR QUAD2.

```
        SUBROUTINE GLAGQ(N,A,B,P,W)
```

THIS SUBROUTINE IS DESIGNED TO BE USED BY THE ROUTINE QUAD FOR APPROXIMATING THE VALUE OF THE IMPROPER INTEGRAL I(F;A,B) OF F(X)*EXP(-X) OVER THE SEMI-INFINITE INTERVAL (A, + OR -VE INFINITY) USING AN N-POINT GAUSS-LAGUERRE FORMULA.

THE PURPOSE OF GLAGQ IS TO SUPPLY, TO QUAD, THE N QUADRATURE POINTS AND WEIGHTS FOR INTEGRATING OVER (A, + OF -VE INFINITY). THE ROUTINE DOES THIS BY TAKING THE N POINTS AND WEIGHTS FOR THE "NORMALIZED" INTERVAL (O, +VE INFINITY) AND TRANSFORMING THEM TO THE GIVEN INTERVAL (A, + OR -VE INFINITY).

THIS ROUTINE CAN ALSO BE USED BY THE ROUTINE QUAD2 TO GENERATE PRODUCT GAUSS-LAGUERRE RULES FOR APPROXIMATING INTEGRALS IN 2 DIMENSIONS.

CALLING SEQUENCE: CALL GLAGQ(N,A,B,P,W)

PARAMETERS:

ON ENTRY:

      N   - INTEGER
             THE NUMBER OF QUADRATURE POINTS TO BE USED.
             RESTRICTION: 2 <= N <= 10.

      A   - REAL
             THE FINITE ENDPOINT OF THE INTERVAL OF INTEGRATION.

      B   - REAL
             THE INFINITE ENDPOINT OF THE INTERVAL OF INTEGRATION.
             = 1   - +VE INFINITY,
             = -1   - -VE INFINITY.

ON RETURN:

      P   - REAL(N)
             THE POINTS FOR THE N-POINT GAUSS-LAGUERRE QUADRATURE RULE ON THE INTERVAL (A,B).

      W   - REAL(N)
             THE WEIGHTS FOR THE N-POINT GAUSS-LAGUERRE QUADRATURE RULE ON THE INTERVAL (A,B).

SAMPLE CALLING PROGRAM:
    SEE EITHER OF THE ROUTINES QUAD OR QUAD2.

```
      SUBROUTINE IVODE(FXY,N,XO,YO,XEND,TOL,IFMLA,IND,CMN,LIST)
```

THIS SUBROUTINE COMPUTES A NUMERICAL SOLUTION, AT THE POINT X = XEND, OF THE INITIAL-VALUE PROBLEM FOR A SYSTEM OF N (<=10) FIRST-ORDER ORDINARY DIFFERENTIAL EQUATIONS

$$Y'(X) = F(X,Y(X)), \qquad Y(XO) = YO.$$

THE ROUTINE PERFORMS A STEP-BY-STEP INTEGRATION FROM XO TO XEND USING A SINGLE-STEP NUMERICAL FORMULA. THE PARTICULAR FORMULA TO BE USED IS SPECIFIED BY THE PARAMETER IFMLA.

THE STEP-SIZE USED AT EACH STAGE OF THE INTEGRATION IS DETERMINED BY A STEP-SIZE CHOOSING ALGORITHM THAT WORKS IN THE FOLLOWING WAY. GIVEN A TRIAL STEP-SIZE H, AN INTEGRATION STEP FROM THE POINT XS TO XS + H WILL BE "SUCCESSFUL" IF

$$EST = MAXNM(ERREST) < TOL*(1 + MAXNM(USP1)) = RTOL,$$

WHERE
  MAXNM(.) DENOTES THE MAXIMUM VECTOR NORM,
  USP1 IS THE NUMERICAL SOLUTION, COMPUTED BY THE FORMULA,
    AT THE POINT XSP1 = XS + H,
  ERREST IS THE VECTOR OF ESTIMATES FOR THE LOCAL ERROR IN
    EACH COMPONENT OF USP1, AND
  TOL IS A PRESCRIBED TOLERANCE FOR THE LOCAL ERRORS.
IF THE STEP IS UNSUCCESSFUL, THAT IS, THE ABOVE TEST FAILS, A NEW (AND SMALLER) TRIAL STEP-SIZE IS COMPUTED AND THE STEP IS REPEATED. THE FORMULA FOR COMPUTING THE NEW STEP-SIZE IS

$$H = FAC*H*(RTOL/ERR)^{1/KP1}$$

WHERE
  KP1 = K+1, K BEING THE ORDER OF ACCURACY OF THE NUMERICAL
    FORMULA, AND
  FAC IS AN ARBITRARY MULTIPLICATIVE FACTOR.
THE PROCESS CONTINUES UNTIL EITHER A STEP IS SUCCESSFUL OR ELSE H < HMIN, WHERE HMIN IS THE MINIMUM PERMISSIBLE STEP-SIZE. THE ROUTINE WILL EXIT WHENEVER H < HMIN. (THIS ABNORMAL RETURN IS INDICATED BY THE PARAMETER IND.) WHEN A STEP IS SUCCESSFUL, A NEW (AND LARGER) STEP-SIZE IS COMPUTED FROM THE SAME FORMULA. THE MINIMUM OF THIS VALUE AND HMAX, THE MAXIMUM PERMISSIBLE STEP-SIZE, IS USED AS AN INITIAL TRIAL STEP-SIZE FOR THE NEXT STEP, AND SO ON UNTIL THE ENDPOINT XEND IS REACHED.

THERE ARE SEVERAL CONTROL PARAMETERS THAT CAN AFFECT THE PERFORMANCE OF THE ALGORITHM. THE ROUTINE PROVIDES THE THE USER WITH THE OPTION OF EITHER SPECIFYING VALUES FOR THEM OR ELSE USING DEFAULT VALUES. SEE THE DESCRIPTIONS OF THE PARAMETERS IND AND CMN FOR DETAILS.

CALLING SEQUENCE: CALL IVODE(FXY,N,X0,Y0,XEND,TOL,IFMLA,
                               IND,CMN,LIST)

PARAMETERS:

ON ENTRY:

    FXY   - SUBROUTINE NAME
            USER-SUPPLIED SUBROUTINE, FXY(N,X,Y,YPRIME), FOR EVALUATING THE RIGHT HAND SIDE F(X,Y) OF THE DIFFERENTIAL SYSTEM (RETURNED IN YPRIME).

    N   - INTEGER
            THE NUMBER OF EQUATIONS IN THE SYSTEM.
            N MUST BE <= 10.

    X0   - REAL
            INITIAL VALUE OF THE INDEPENDENT VARIABLE X.

    Y0   - REAL(N)
            INITIAL VALUES OF THE DEPENDENT VARIABLES $Y(X)$: $Y0(J) = Y_J(X0)$.

    XEND   - REAL
            THE VALUE OF THE INDEPENDENT VARIABLE AT WHICH THE SOLUTION Y(XEND) IS DESIRED.

    TOL   - REAL
            TOLERANCE FOR THE LOCAL ERROR (PER STEP).

    IFMLA   - INTEGER
            A PARAMETER FOR SPECIFYING THE NUMERICAL FORMULA TO BE USED. THE FORMULA MUST BE IMPLEMENTED AS A SUBROUTINE FOR TAKING A SINGLE INTEGRATION STEP AND RETURNING THE COMPUTED SOLUTION AS WELL AS AN ESTIMATE FOR ITS LOCAL ERROR. THE FOLLOWING VALUES OF IFMLA WILL ACCESS THE FORMULAS THAT HAVE BEEN IMPLEMENTED IN THE PACKAGE.

            EXPLICIT FORMULAS:
            = 1 - RKF4 - RUNGE-KUTTA-FEHLBERG FOURTH-ORDER FORMULA;
            = 2 - EULER - EULER'S FORMULA;

            IMPLICIT FORMULAS:
            = 31 - TZN   - TRAPEZOIDAL RULE FORMULA WITH NEWTON ITERATION;
            = 32 - TZNF - SAME AS TZN BUT WITH THE JACOBIAN FIXED;
            = 33 - TZMN - TRAPEZOIDAL RULE FORMULA WITH MOD. NEWTON ITERATION

```
                        - FINITE DIFFERENCE APPROX.
                          FOR THE JACOBIAN;
            = 34 - TZMNF - SAME  AS  TZMN BUT WITH THE
                          APPROX. JACOBIAN FIXED.

        FOR MORE INFORMATION, SEE  THE  DOCUMENTATION
        FOR THE APPROPRIATE SUBROUTINE.  IN ADDITION,
        A USER-SUPPLIED FORMULA CAN  BE  CALLED  BY
        SETTING IFMLA

            = 4  - FMLA  - USER-SUPPLIED FORMULA.

        THE  RESTRICTIONS  ARE  THAT  IT  MUST  BE  A
        ONE-STEP  FORMULA  AND  THAT  ITS  SUBROUTINE
        IMPLEMENTATION MUST CONFORM TO THE FOLLOWING
        CALLING SEQUENCE:
            CALL FMLA(FXY,N,XS,US,USP1,ERREST,CMN)
        WHERE THE PARAMETERS ARE THE SAME AS FOR  THE
        SUBROUTINE  TZN.  IF THIS OPTION IS NOT USED,
        A  DUMMY  SUBROUTINE  OF  THE  FORM GIVEN IN THE
        SAMPLE CALLING PROGRAMS SHOULD BE PROVIDED IN
        ORDER TO AVOID COMPILER WARNING MESSAGES.

IND     - INTEGER
        A PARAMETER FOR INDICATING WHETHER OR NOT THE
        USER WISHES TO USE THE  OPTION  OF  SPECIFYING
        ANY OF THE CONTROL PARAMETERS IN CMN.
            = 0 -  DEFAULT  VALUES  FOR  EACH PARAMETER
                   ARE TO BE USED;
            = 1 -  SOME OR ALL VALUES ARE TO  BE  USER-
                   SUPPLIED.

CMN     - REAL(12)
        COMMUNICATIONS VECTOR.  COMPONENTS 1 - 10  OF
        CMN  CORRESPOND TO CONTROL PARAMETERS USED BY
        THE SUBROUTINE.  THE  USER  MAY  SUPPLY  VALUES
        FOR ANY NUMBER OF THEM AND USE DEFAULT VALUES
        FOR  THE  REMAINING ONES.  IF IND=1, A DEFAULT
        VALUE FOR THE J-TH COMPONENT CAN BE REQUESTED
        BY  SETTING  CMN(J) = 0.E0.  FOLLOWING  IS  A
        LIST OF THE DEFINITION OF EACH COMPONENT  AND
        ITS DEFAULT VALUE:
                 DEFINITION              DEFAULT
         CMN(1) - HMIN              10*EPS*MAX(|X0|,1.0)
            (2) - HSTART            HMAX*TOL**(1/KP1)
            (3) - HMAX              2.0
            (4) - FAC               0.9
            (5) - KP1               DEPENDS ON IFMLA
            (6) - ITMAX             2
            (7) - XTOL              4*EPS
            (8) - FTOL              4*EPS
            (9) - ZERO              1.E-30
                  THRESHOLD
           (10) - DELTA (FOR        SQRT(EPS)
                  FIN. DIFF.)
```

WHERE EPS IS MACHINE EPSILON.
CMN(2) IS USED SUBSEQUENTLY TO RECORD THE CURRENT STEP-SIZE H.
CMN(5) NEED ONLY BE SPECIFIED IN THE CASE OF A USER-SUPPLIED FORMULA, THAT IS, WHEN IFMLA=4.
CMN(6), CMN(7) AND CMN(8) ARE CONTROL PARAMETERS FOR SOLVING A NONLINEAR SYSTEM OF EQUATIONS. THEY ARE USED ONLY IN THE CASE OF AN IMPLICIT FORMULA.
CMN(9) IS FOR PROTECTION AGAINST UNDERFLOW. EACH OF THE SUBROUTINES REFERENCED BY IFMLA INCLUDES THE TEST, FOR 1<=J<=N,
    IF (¦Y(J)¦ <= CMN(9)) THEN
       Y(J) = 0.E0
CMN(10) IS THE AMOUNT OF THE SHIFT USED FOR COMPUTING FINITE DIFFERENCES (IN TZMN AND TZMNF).
CMN(11) AND CMN(12) ARE FOR RETURNING INFORMATION TO THE USER.

LIST   - INTEGER
      A PARAMETER FOR CONTROLLING THE OUTPUT OF INTERMEDIATE RESULTS.
      = 0 - NO OUTPUT;
      = K - OUTPUT EVERY K SUCCESSFUL STEPS.

ON RETURN:

XO   - THE FINAL VALUE OF THE DEPENDENT VARIABLE:
       = XEND FOR NORMAL RETURN;
       ^= XEND FOR ABNORMAL RETURN.

YO   - THE NUMERICAL SOLUTION AT THE RETURNED XO.

IND   - INDICATOR OF SUCCESS OR FAILURE:
      = 1 - NORMAL RETURN. IN ORDER TO CONTINUE THE INTEGRATION USING THE CURRENT VALUES OF THE CONTROL PARAMETERS, SIMPLY RE-CALL IVODE WITH THE NEW XEND.
      = -1 - ABNORMAL RETURN. INTEGRATION COULD NOT BE COMPLETED BECAUSE H < HMIN.
      = -2 - ABNORMAL RETURN. INTEGRATION COULD NOT BE COMPLETED BECAUSE THE JACOBIAN (OR ITS APPROXIMATION) IS SINGULAR -- IMPLICIT FORMULAS ONLY.

CMN   - CMN(2) - SUGGESTED VALUE OF HSTART IF THE INTEGRATION IS TO BE CONTINUED BEYOND XEND.
      (11) - INDICATES STATUS OF JACOBIAN
          = 0.E0 - NONSINGULAR
          = 2.E0 - SINGULAR

(12) - GIVES THE NUMBER OF STEPS TAKEN TO COMPLETE THE INTEGRATION.
THE REMAINING ENTRIES CONTAIN THE CURRENT VALUES OF THE CONTROL PARAMETERS.

SAMPLE CALLING PROGRAM: SEE ANY OF THE SUBROUTINES RKF4, TZN, TZNF, TZMN, OR TZMNF.

        SUBROUTINE RKF4(FXY,N,XS,US,USP1,ERREST,CMN)

        THIS SUBROUTINE PERFORMS A SINGLE STEP INTEGRATION OF THE
        INITIAL-VALUE PROBLEM FOR THE SYSTEM OF N (<=10) FIRST-
        ORDER ORDINARY DIFFERENTIAL EQUATIONS

                Y'(X) = F(X,Y(X)),   Y(XS) = US.

        THE RUNGE-KUTTA-FEHLBERG 4TH-ORDER FORMULA IS USED.

        CALLING SEQUENCE: CALL RKF4(FXY,N,XS,US,USP1,ERREST,CMN)

        PARAMETERS:

        ON ENTRY:

                FXY   - SUBROUTINE NAME
                          USER-SUPPLIED SUBROUTINE, FXY(N,X,Y,YPRIME),
                          FOR EVALUATING THE RIGHT HAND SIDE F(X,Y)
                          OF THE DIFFERENTIAL SYSTEM (RETURNED IN
                          YPRIME).

                N   - INTEGER
                          THE NUMBER OF EQUATIONS IN THE SYSTEM.
                          N MUST BE <= 10.

                XS   - REAL
                          INITIAL VALUE OF THE INDEPENDENT VARIABLE X.

                US   - REAL(N)
                          INITIAL VALUES OF THE DEPENDENT VARIABLES
                          Y(X).

                CMN   - REAL(12)
                          COMMUNICATIONS VECTOR. FOR DEFINITIONS OF
                          ITS COMPONENTS, SEE THE DOCUMENTATION FOR THE
                          SUBROUTINE IVODE.

        ON RETURN:

                USP1   - REAL(N)
                          THE COMPUTED VALUES OF THE SOLUTION AT
                          XSP1 = XS + H.

                ERREST - REAL(N)
                          ESTIMATES FOR THE ERROR IN EACH COMPONENT OF
                          USP1.

        SAMPLE CALLING PROGRAM:

        C   ****************

```
C      *
C      * THIS PROGRAM USES RKF4 WITH THE   ROUTINE  IVODE.  THE
C      * INITIAL   STEP-SIZE   HSTART   IS SET EQUAL TO   0.01 AND
C      * THE MAXIMUM STEP-SIZE HMAX IS TO BE 0.5.  ALL OF   THE
C      * OTHER  CONTROL  PARAMETERS  USED   BY  IVODE ARE TO BE
C      * DEFAULT VALUES.  THE   PROGRAM  COMPUTES  A  NUMERICAL
C      * SOLUTION AT EACH OF THE POINTS X = 0.5, 1.0,..., 5.0.
C      *
C      ***************
       REAL X0,Y0(10),XEND,CMN(12),TOL
       EXTERNAL FXY
       READ (5,*) N,X0,(Y0(J),J=1,N)
       TOL = 1.E-5
       IFMLA = 1
       IND = 1
       LIST = 0
       DO 10 I = 1,10
          CMN(I) = 0.E0
   10  CONTINUE
       CMN(2) = 1.E-2
       CMN(3) = 0.5E0
       DO 20 K = 1,10
          XEND = FLOAT(K)*0.5E0
          CALL IVODE(FXY,N,X0,Y0,XEND,TOL,IFMLA,IND,CMN,LIST)
             IF (IND .LT. 0) GO TO 30
          WRITE (6,*) 'SOLUTION AT X = ',X0
          WRITE (6,*) (Y0(J),J=1,N)
          NSTP = IFIX(CMN(12))
          WRITE (6,*) '# STEPS = ',NSTP
   20  CONTINUE
       STOP
   30  WRITE (6,*) 'INTEGRATION FAILED : IND = ',IND
       STOP
       END

       SUBROUTINE FXY(N,X,Y,YPRIME)
       REAL X,Y(N),YPRIME(N)
        ...    EXPRESSIONS TO EVALUATE EACH COMPONENT OF F(X,Y)
                YPRIME(J) = F (X,Y),     1<=J<=N.
                             J
       RETURN
       END

       SUBROUTINE FMLA(FXY,N,XS,US,USP1,ERREST,CMN)
       REAL XS,US(N),USP1(N),ERREST(N),CMN(12)
       EXTERNAL FFU,JFFU
       RETURN
       END
```

```
      SUBROUTINE EULER(FXY,N,XS,US,USP1,ERREST,CMN)
```

THIS SUBROUTINE PERFORMS A SINGLE STEP INTEGRATION OF THE INITIAL-VALUE PROBLEM FOR THE SYSTEM OF N (<=10) FIRST-ORDER ORDINARY DIFFERENTIAL EQUATIONS

$$Y'(X) = F(X,Y(X)), \qquad Y(XS) = US.$$

EULER'S FORMULA IS USED.

****
   NOTE : SINCE EULER'S FORMULA IS NOT VERY ACCURATE,
          USE OF THIS ROUTINE FOR ANYTHING BUT
          INSTRUCTIVE PURPOSES IS NOT RECOMMENDED.
****

CALLING SEQUENCE: CALL EULER(FXY,N,XS,US,USP1,ERREST,CMN)

PARAMETERS:

ON ENTRY:

    FXY   - SUBROUTINE NAME
              USER-SUPPLIED SUBROUTINE, FXY(N,X,Y,YPRIME),
              FOR EVALUATING THE RIGHT HAND SIDE F(X,Y)
              OF THE DIFFERENTIAL SYSTEM (RETURNED IN
              YPRIME).

    N   - INTEGER
              THE NUMBER OF EQUATIONS IN THE SYSTEM.
              N MUST BE <= 10.

    XS   - REAL
              INITIAL VALUE OF THE INDEPENDENT VARIABLE X.

    US   - REAL(N)
              INITIAL VALUES OF THE DEPENDENT VARIABLES
              Y(X).

    CMN   - REAL(12)
              COMMUNICATIONS VECTOR. FOR DEFINITIONS OF
              ITS COMPONENTS, SEE THE DOCUMENTATION FOR THE
              SUBROUTINE IVODE.

ON RETURN:

    USP1   - REAL(N)
              THE COMPUTED VALUES OF THE SOLUTION AT
              XSP1 = XS + H.

    ERREST - REAL(N)
              ESTIMATES FOR THE ERROR IN EACH COMPONENT OF

USP1.

SAMPLE CALLING PROGRAM:

```
C      ***************
C      *
C      * THIS PROGRAM USES A FIXED STEP-SIZE H.  THE ONLY
C      * ELEMENTS OF THE COMMUNICATIONS VECTOR CMN THAT ARE
C      * USED BY EULER ARE THE STEP-SIZE H = CMN(2) AND THE
C      * ZERO THRESHOLD = CMN(9) (FOR UNDERFLOW PROTECTION).
C      *
C      ***************
       REAL X0,Y0(10),XEND,YPH(10),ERREST(10),CMN(12),TOL
       EXTERNAL FXY
       READ (5,*) N,X0,XEND,H,(Y0(J),J=1,N)
       CMN(2) = H
       CMN(9) = 1.E-30
    10 IF (X0+H .GT. XEND) GO TO 20
       CALL EULER(FXY,N,X0,Y0,YPH,ERREST,CMN)
       X0 = X0 + H
       DO 15 J = 1,N
           Y0(J) = YPH(J)
    15 CONTINUE
    20 WRITE (6,*) 'SOLUTION AT X = ',X0
       DO 25 J = 1,N
           WRITE (6,*) J,Y0(J)
    25 CONTINUE
       STOP
       END

       SUBROUTINE FXY(N,X,Y,YPRIME)
       REAL X,Y(N),YPRIME(N)
       ...  EXPRESSIONS TO EVALUATE EACH COMPONENT OF F(X,Y)
               YPRIME(J) = F (X,Y),       1<=J<=N.
                            J
       RETURN
       END
```

```
      SUBROUTINE TZN(FXY,N,XS,US,USP1,ERREST,CMN)
```

THIS SUBROUTINE PERFORMS A SINGLE STEP INTEGRATION OF THE INITIAL-VALUE PROBLEM FOR THE SYSTEM OF N (<=10) FIRST-ORDER ORDINARY DIFFERENTIAL EQUATIONS

    Y'(X) = F(X,Y(X)),   Y(XS) = US.

THE TRAPEZOIDAL RULE FORMULA IS USED, THAT IS, THE NUMERICAL SOLUTION USP1, AT XSP1 = XS + H, IS DEFINED BY THE FORMULA

    USP1 = US + 0.5*H*(FS + FSP1),

WHERE FS = F(XS,US) AND SIMILARLY FOR FSP1. SINCE THIS IS AN IMPLICIT FORMULA, USP1 MUST BE COMPUTED BY SOLVING THE SYSTEM OF N NONLINEAR EQUATIONS

    FF(USP1) = USP1 - 0.5*H*F(XSP1,USP1) - C = 0,

WHERE C = US + 0.5*H*FS IS KNOWN. NEWTON'S METHOD IS USED TO SOLVE THIS SYSTEM.

THE USER MUST SUPPLY SUBROUTINES TO EVALUATE BOTH F(X,Y) AND ITS JACOBIAN MATRIX. THEY MUST CONFORM TO THE CALLING SEQUENCES FXY(N,X,Y,YPRIME) AND JFXY(N,X,Y,AJFXY), RESPECTIVELY. (SEE THE SAMPLE CALLING PROGRAM.)

THE ERROR ESTIMATE RETURNED BY THE ROUTINE IS COMPUTED BY THE ONE STEP, TWO HALF-STEPS PROCEDURE.

THE ROUTINE CALLS THE NEWTON SUBROUTINE SNWTN. THE UTILITY ROUTINES FFU AND JFFU ARE USED TO FACILLITATE EVALUATION, BY SNWTN, OF THE FUNCTION FF(U) AND ITS JACOBIAN MATRIX. SOME OF THE INFORMATION REQUIRED BY FFU AND JFFU IS PASSED THROUGH A COMMON BLOCK LABELLED DTPZ.

CALLING SEQUENCE: CALL TZN(FXY,N,XS,US,USP1,ERREST,CMN)

PARAMETERS:

ON ENTRY:

        FXY   - SUBROUTINE NAME
                  USER-SUPPLIED SUBROUTINE, FXY(N,X,Y,YPRIME),
                  FOR EVALUATING THE RIGHT HAND SIDE F(X,Y)
                  OF THE DIFFERENTIAL SYSTEM (RETURNED IN
                  YPRIME).

        N   - INTEGER
                  THE NUMBER OF EQUATIONS IN THE SYSTEM.
                  N MUST BE <= 10.

XS   - REAL
          INITIAL VALUE OF THE INDEPENDENT VARIABLE X.

US   - REAL(N)
          INITIAL VALUES OF THE DEPENDENT VARIABLES
          Y(X).

CMN   - REAL(12)
          COMMUNICATIONS VECTOR. FOR DEFINITIONS OF
          ITS COMPONENTS, SEE THE DOCUMENTATION FOR THE
          SUBROUTINE IVODE.

ON RETURN:

USP1   - REAL(N)
          THE COMPUTED VALUES OF THE SOLUTION AT
          XSP1 = XS + H.

ERREST - REAL(N)
          ESTIMATES FOR THE ERROR IN EACH COMPONENT OF
          USP1.

SAMPLE CALLING PROGRAM:

```
C     ****************
C     *
C     * THIS PROGRAM USES TZN WITH THE ROUTINE IVODE. THE
C     * INITIAL STEP-SIZE HSTART IS SET EQUAL TO 0.01.   ALL
C     * OF THE OTHER CONTROL PARAMETERS USED BY IVODE ARE TO
C     * BE DEFAULT VALUES. THE PROGRAM COMPUTES A NUMERICAL
C     * SOLUTION AT EACH OF THE POINTS X = 0.5, 1.0,..., 5.0.
C     *
C     ****************
      REAL X0,Y0(10),XEND,CMN(12),TOL
      EXTERNAL FXY
      READ (5,*) N,X0,(Y0(J),J=1,N)
      TOL = 1.E-5
      IFMLA = 31
      IND = 1
      LIST = 0
      DO 10 I = 1,10
         CMN(I) = 0.E0
   10 CONTINUE
      CMN(2) = 1.E-2
      DO 20 K = 1,10
         XEND = FLOAT(K)*0.5E0
         CALL IVODE(FXY,N,X0,Y0,XEND,TOL,IFMLA,IND,CMN,LIST)
            IF (IND .LT. 0) GO TO 30
         WRITE (6,*) 'SOLUTION AT X = ',X0
         WRITE (6,*) (Y0(J),J=1,N)
         NSTP = IFIX(CMN(12))
         WRITE (6,*) '# STEPS = ',NSTP
```

```
   20 CONTINUE
      STOP
   30 WRITE (6,*) 'INTEGRATION FAILED : IND = ',IND
      STOP
      END

      SUBROUTINE FXY(N,X,Y,YPRIME)
      REAL X,Y(N),YPRIME(N)
       ...   EXPRESSIONS TO EVALUATE EACH COMPONENT OF F(X,Y)
              YPRIME(J) = F (X,Y),     1<=J<=N.
                            J
      RETURN
      END

      SUBROUTINE JFXY(N,X,Y,AJFXY)
      REAL X,Y(N),AJFXY(N,N)
       ...   EXPRESSIONS TO EVALUATE EACH COMPONENT OF THE
             JACOBIAN MATRIX AJFXY --
                AJFXY(I,J) = DF (X,Y) / DY (X)
                               I         J
      RETURN
      END

      SUBROUTINE FMLA(FXY,N,XS,US,USP1,ERREST,CMN)
      REAL XS,US(N),USP1(N),ERREST(N),CMN(12)
      EXTERNAL FFU,JFFU
      RETURN
      END
```

```
      SUBROUTINE TZNF(FXY,N,XS,US,USP1,ERREST,CMN)
```

THIS SUBROUTINE PERFORMS A SINGLE STEP INTEGRATION OF THE INITIAL-VALUE PROBLEM FOR THE SYSTEM OF N (<=10) FIRST-ORDER ORDINARY DIFFERENTIAL EQUATIONS

     Y'(X) = F(X,Y(X)),   Y(XS) = US.

THE TRAPEZOIDAL RULE FORMULA IS USED, THAT IS, THE NUMERICAL SOLUTION USP1, AT XSP1 = XS + H, IS DEFINED BY THE FORMULA

    USP1 = US + 0.5*H*(FS + FSP1),

WHERE FS = F(XS,US) AND SIMILARLY FOR FSP1. SINCE THIS IS AN IMPLICIT FORMULA, USP1 MUST BE COMPUTED BY SOLVING THE SYSTEM OF N NONLINEAR EQUATIONS

    FF(USP1) = USP1 - 0.5*H*F(XSP1,USP1) - C = 0,

WHERE C = US + 0.5*H*FS IS KNOWN. NEWTON'S METHOD, WITH THE JACOBIAN FIXED AFTER THE FIRST ITERATION, IS USED TO SOLVE THIS SYSTEM.

THE USER MUST SUPPLY SUBROUTINES TO EVALUATE BOTH F(X,Y) AND ITS JACOBIAN MATRIX. THEY MUST CONFORM TO THE CALLING SEQUENCES FXY(N,X,Y,YPRIME) AND JFXY(N,X,Y,AJFXY), RESPECTIVELY. (SEE THE SAMPLE CALLING PROGRAM.)

THE ERROR ESTIMATE RETURNED BY THE ROUTINE IS COMPUTED BY THE ONE STEP, TWO HALF-STEPS PROCEDURE.

FOR CONVENIENCE, THE ROUTINE USES THE UTILITY ROUTINES FFU AND JFFU TO EVALUATE FF(U) AND ITS JACOBIAN MATRIX. SOME OF THE INFORMATION REQUIRED BY THESE ROUTINES IS PASSED THROUGH A COMMON BLOCK LABELLED DTPZ.

CALLING SEQUENCE: CALL TZNF(FXY,N,XS,US,USP1,ERREST,CMN)

PARAMETERS:

ON ENTRY:

    FXY   - SUBROUTINE NAME
               USER-SUPPLIED SUBROUTINE, FXY(N,X,Y,YPRIME), FOR EVALUATING THE RIGHT HAND SIDE F(X,Y) OF THE DIFFERENTIAL SYSTEM (RETURNED IN YPRIME).

    N   - INTEGER
               THE NUMBER OF EQUATIONS IN THE SYSTEM.
               N MUST BE <= 10.

```
        XS      - REAL
                  INITIAL VALUE OF THE INDEPENDENT VARIABLE X.

        US      - REAL(N)
                  INITIAL VALUES OF THE DEPENDENT VARIABLES
                  Y(X).

        CMN     - REAL(12)
                  COMMUMICATIONS VECTOR.  FOR DEFINITIONS OF
                  ITS COMPONENTS, SEE THE DOCUMENTATION FOR THE
                  SUBROUTINE IVODE.

ON RETURN:

        USP1    - REAL(N)
                  THE COMPUTED VALUES OF THE SOLUTION AT
                  XSP1 = XS + H.

        ERREST  - REAL(N)
                  ESTIMATES  FOR THE ERROR IN EACH COMPONENT OF
                  USP1.


SAMPLE CALLING PROGRAM:

C       ***************
C       *
C       * THIS PROGRAM USES TZNF WITH THE ROUTINE  IVODE.   ALL
C       * OF   THE CONTROL PARAMETERS USED BY IVODE  ARE  TO  BE
C       * DEFAULT  VALUES.   THE  PROGRAM  COMPUTES A NUMERICAL
C       * SOLUTION AT EACH OF THE POINTS X = 1.0, 2.0,..., 10.0.
C       *
C       ***************
        REAL X0,Y0(10),XEND,CMN(12),TOL
        EXTERNAL FXY
        READ (5,*) N,X0,(Y0(J),J=1,N)
        TOL = 1.E-5
        IFMLA = 32
        IND = 0
        LIST = 0
        DO 20 K = 1,10
           XEND = FLOAT(K)
           CALL IVODE(FXY,N,X0,Y0,XEND,TOL,IFMLA,IND,CMN,LIST)
              IF (IND .LT. 0) GO TO 30
           WRITE (6,*) 'SOLUTION AT X = ',X0
           WRITE (6,*) (Y0(J),J=1,N)
           NSTP = IFIX(CMN(12))
           WRITE (6,*) '# STEPS = ',NSTP
    20  CONTINUE
        STOP
    30  WRITE (6,*) 'INTEGRATION FAILED : IND = ',IND
        STOP
        END
```

```
      SUBROUTINE FXY(N,X,Y,YPRIME)
      REAL X,Y(N),YPRIME(N)
  ...    EXPRESSIONS TO EVALUATE EACH COMPONENT OF F(X,Y)
             YPRIME(J) = F (X,Y),      1<=J<=N.
                          J
      RETURN
      END

      SUBROUTINE JFXY(N,X,Y,AJFXY)
      REAL X,Y(N),AJFXY(N,N)
  ...    EXPRESSIONS TO EVALUATE EACH COMPONENT OF THE
         JACOBIAN MATRIX AJFXY --
             AJFXY(I,J) = DF (X,Y) / DY (X)
                            I          J
      RETURN
      END

      SUBROUTINE FMLA(FXY,N,XS,US,USP1,ERREST,CMN)
      REAL XS,US(N),USP1(N),ERREST(N),CMN(12)
      EXTERNAL FFU,JFFU
      RETURN
      END
```

```
      SUBROUTINE TZMN(FXY,N,XS,US,USP1,ERREST,CMN)
```

THIS SUBROUTINE PERFORMS A SINGLE STEP INTEGRATION OF THE INITIAL-VALUE PROBLEM FOR THE SYSTEM OF N (<=10) FIRST-ORDER ORDINARY DIFFERENTIAL EQUATIONS

$$Y'(X) = F(X,Y(X)), \qquad Y(XS) = US.$$

THE TRAPEZOIDAL RULE FORMULA IS USED, THAT IS, THE NUMERICAL SOLUTION USP1, AT XSP1 = XS + H, IS DEFINED BY THE FORMULA

$$USP1 = US + 0.5*H*(FS + FSP1),$$

WHERE FS = F(XS,US) AND SIMILARLY FOR FSP1. SINCE THIS IS AN IMPLICIT FORMULA, USP1 MUST BE COMPUTED BY SOLVING THE SYSTEM OF N NONLINEAR EQUATIONS

$$FF(USP1) = USP1 - 0.5*H*F(XSP1,USP1) - C = 0,$$

WHERE C = US + 0.5*H*FS IS KNOWN. A MODIFIED NEWTON'S METHOD -- USING A FINITE DIFFERENCE APPROXIMATION FOR THE JACOBIAN MATRIX -- IS USED TO SOLVE THIS SYSTEM.

THE ERROR ESTIMATE RETURNED BY THE ROUTINE IS COMPUTED BY THE ONE STEP, TWO HALF-STEPS PROCEDURE.

FOR CONVENIENCE, THE ROUTINE USES THE UTILITY ROUTINE FFU TO EVALUATE FF(U). SOME OF THE INFORMATION REQUIRED BY FFU IS PASSED THROUGH A COMMON BLOCK LABELLED DTPZ.

CALLING SEQUENCE: CALL TZMN(FXY,N,XS,US,USP1,ERREST,CMN)

PARAMETERS:

ON ENTRY:

    FXY   - SUBROUTINE NAME
              USER-SUPPLIED SUBROUTINE, FXY(N,X,Y,YPRIME), FOR EVALUATING THE RIGHT HAND SIDE F(X,Y) OF THE DIFFERENTIAL SYSTEM (RETURNED IN YPRIME).

    N   - INTEGER
              THE NUMBER OF EQUATIONS IN THE SYSTEM.
              N MUST BE <= 10.

    XS   - REAL
              INITIAL VALUE OF THE INDEPENDENT VARIABLE X.

    US   - REAL(N)
              INITIAL VALUES OF THE DEPENDENT VARIABLES

                        Y(X).

        CMN   - REAL(12)
                  COMMUNICATIONS VECTOR. FOR DEFINITIONS OF
                  ITS COMPONENTS, SEE THE DOCUMENTATION FOR THE
                  SUBROUTINE IVODE.

ON RETURN:

        USP1   - REAL(N)
                  THE COMPUTED VALUES OF THE SOLUTION AT
                  XSP1 = XS + H.

        ERREST - REAL(N)
                  ESTIMATES FOR THE ERROR IN EACH COMPONENT OF
                  USP1.

SAMPLE CALLING PROGRAM:

```
C     ***************
C     *
C     * THIS PROGRAM USES TZMN WITH THE ROUTINE   IVODE.   THE
C     * MAXIMUM  STEP-SIZE  HMAX  IS SET EQUAL TO 0.5 AND THE
C     * MAXIMUM NUMBER OF (MODIFIED  NEWTON)  ITERATIONS  PER
C     * STEP  IS 3.  ALL OF THE OTHER CONTROL PARAMETERS USED
C     * BY IVODE ARE  TO  BE  DEFAULT  VALUES.    THE  PROGRAM
C     * COMPUTES  A  NUMERICAL SOLUTION AT EACH OF THE POINTS
C     * X = 0.5, 1.0,..., 5.0.
C     *
C     ***************
      REAL X0,Y0(10),XEND,CMN(12),TOL
      EXTERNAL FXY
      READ (5,*) N,X0,(Y0(J),J=1,N)
      TOL = 1.E-5
      IFMLA = 33
      IND = 1
      LIST = 0
      DO 10 I = 1,10
         CMN(I) = 0.E0
   10 CONTINUE
      CMN(3) = 0.5E0
      CMN(6) = 3
      DO 20 K = 1,10
         XEND = FLOAT(K)*0.5E0
         CALL IVODE(FXY,N,X0,Y0,XEND,TOL,IFMLA,IND,CMN,LIST)
            IF (IND .LT. 0) GO TO 30
         WRITE (6,*) 'SOLUTION AT X = ',X0
         WRITE (6,*) (Y0(J),J=1,N)
         NSTP = IFIX(CMN(12))
         WRITE (6,*) '# STEPS = ',NSTP
   20 CONTINUE
      STOP
   30 WRITE (6,*) 'INTEGRATION FAILED : IND = ',IND
```

```
      STOP
      END

      SUBROUTINE FXY(N,X,Y,YPRIME)
      REAL X,Y(N),YPRIME(N)
       ...   EXPRESSIONS TO EVALUATE EACH COMPONENT OF F(X,Y)
             YPRIME(J) = F (X,Y),      1<=J<=N.
                          J
      RETURN
      END

      SUBROUTINE FMLA(FXY,N,XS,US,USP1,ERREST,CMN)
      REAL XS,US(N),USP1(N),ERREST(N),CMN(12)
      EXTERNAL FFU,JFFU
      RETURN
      END
```

```
    SUBROUTINE TZMNF(FXY,N,XS,US,USP1,ERREST,CMN)
```

THIS SUBROUTINE PERFORMS A SINGLE STEP INTEGRATION OF THE INITIAL-VALUE PROBLEM FOR THE SYSTEM OF N (<=10) FIRST-ORDER ORDINARY DIFFERENTIAL EQUATIONS

```
        Y'(X) = F(X,Y(X)),      Y(XS) = US.
```

THE TRAPEZOIDAL RULE FORMULA IS USED, THAT IS, THE NUMERICAL SOLUTION USP1, AT XSP1 = XS + H, IS DEFINED BY THE FORMULA

```
        USP1 = US + 0.5*H*(FS + FSP1),
```

WHERE FS = F(XS,US) AND SIMILARLY FOR FSP1. SINCE THIS IS AN IMPLICIT FORMULA, USP1 MUST BE COMPUTED BY SOLVING THE SYSTEM OF N NONLINEAR EQUATIONS

```
        FF(USP1) = USP1 - 0.5*H*F(XSP1,USP1) - C = 0,
```

WHERE C = US + 0.5*H*FS IS KNOWN. A MODIFIED NEWTON'S METHOD -- USING A FINITE DIFFERENCE APPROXIMATION FOR THE JACOBIAN MATRIX, FIXED AFTER THE FIRST ITERATION -- IS USED TO SOLVE THIS SYSTEM.

THE ERROR ESTIMATE RETURNED BY THE ROUTINE IS COMPUTED BY THE ONE STEP, TWO HALF-STEPS PROCEDURE.

FOR CONVENIENCE, THE ROUTINE USES THE UTILITY ROUTINE FFU TO EVALUATE FF(U). SOME OF THE INFORMATION REQUIRED BY FFU IS PASSED THROUGH A COMMON BLOCK LABELLED DTPZ.

CALLING SEQUENCE: CALL TZMNF(FXY,N,XS,US,USP1,ERREST,CMN)

PARAMETERS:

ON ENTRY:

    FXY   - SUBROUTINE NAME
              USER-SUPPLIED SUBROUTINE, FXY(N,X,Y,YPRIME), FOR EVALUATING THE RIGHT HAND SIDE F(X,Y) OF THE DIFFERENTIAL SYSTEM (RETURNED IN YPRIME).

    N   - INTEGER
              THE NUMBER OF EQUATIONS IN THE SYSTEM.
              N MUST BE <= 10.

    XS   - REAL
              INITIAL VALUE OF THE INDEPENDENT VARIABLE X.

    US   - REAL(N)

```
                        INITIAL VALUES  OF   THE   DEPENDENT   VARIABLES
                        Y(X).

        CMN     -  REAL(12)
                   COMMUNICATIONS VECTOR.  FOR DEFINITIONS OF
                   ITS COMPONENTS, SEE THE DOCUMENTATION FOR THE
                   SUBROUTINE IVODE.

ON RETURN:

        USP1    -  REAL(N)
                   THE COMPUTED VALUES OF THE SOLUTION AT
                   XSP1 = XS + H.

        ERREST  -  REAL(N)
                   ESTIMATES  FOR THE ERROR IN EACH COMPONENT OF
                   USP1.


SAMPLE CALLING PROGRAM:

C       ***************
C       *
C       * THIS  PROGRAM  USES  TZMNF  WITH   THE    ROUTINE IVODE.
C       * ALL OF THE CONTROL PARAMETERS USED BY  IVODE  ARE   TO
C       * BE DEFAULT VALUES.
C       *
C       ***************
        REAL X0,Y0(10),XEND,CMN(12),TOL
        EXTERNAL FXY
        READ (5,*) N,X0,(Y0(J),J=1,N)
        TOL = 1.E-5
        IFMLA = 34
        IND = 0
        LIST = 0
        CALL IVODE(FXY,N,X0,Y0,XEND,TOL,IFMLA,IND,CMN,LIST)
           IF (IND .LT. 0) GO TO 30
        WRITE (6,*) 'SOLUTION AT X = ',X0
        WRITE (6,*) (Y0(J),J=1,N)
        NSTP = IFIX(CMN(12))
        WRITE (6,*) '# STEPS = ',NSTP
        STOP
     30 WRITE (6,*) 'INTEGRATION FAILED : IND = ',IND
        STOP
        END

        SUBROUTINE FXY(N,X,Y,YPRIME)
        REAL X,Y(N),YPRIME(N)
          ...    EXPRESSIONS TO EVALUATE EACH COMPONENT OF F(X,Y)
                    YPRIME(J) = F (X,Y),      1<=J<=N.
                                 J
        RETURN
        END
```

```
SUBROUTINE FMLA(FXY,N,XS,US,USP1,ERREST,CMN)
REAL XS,US(N),USP1(N),ERREST(N),CMN(12)
EXTERNAL FFU,JFFU
RETURN
END
```

SUBROUTINE FFU(N,U,FF)

THIS IS A UTILITY SUBROUTINE FOR EVALUATING THE FUNCTION FF(U), ARISING IN THE TRAPEZOIDAL RULE FORMULA. ITS PURPOSE IS TO FACILLITATE CALLS, BY THE ROUTINE TZN, TO THE NEWTON SUBROUTINE SNWTN. (FFU IS ALSO CALLED BY THE OTHER TRAPEZOIDAL FORMULA ROUTINES TZNF, TZMN, AND TZMNF.)

SOME OF THE INFORMATION USED BY THIS ROUTINE IS PASSED THROUGH A COMMON BLOCK NAMED DTPZ, THAT IS, WITH THE STATEMENT
    COMMON /DTPZ/HSTP,XSTP,C

          SUBROUTINE JFFU(N,U,AJFF)

THIS IS A UTILITY SUBROUTINE FOR EVALUATING THE JACOBIAN
MATRIX JFF(U) OF THE FUNCTION FF(U), ARISING IN THE
TRAPEZOIDAL RULE FORMULA. ITS PURPOSE IS TO FACILLITATE
CALLS, BY THE SUBROUTINE TZN, TO THE NEWTON SUBROUTINE
SNWTN. (JFFU IS ALSO CALLED BY THE ROUTINE TZNF.)

SOME OF THE INFORMATION USED BY THIS ROUTINE IS PASSED
THROUGH A COMMON BLOCK NAMED DTPZ, THAT IS, WITH THE
STATEMENT
     COMMON /DTPZ/HSTP,XSTP,C

*tty:=teaman.mem/freeff:55